Self-Checking and Fault-Tolerant Digital Design

Self-Checking and Fault-Tolerant Digital Design

Parag K. Lala

University of Arkansas

MORGAN KAUFMANN PUBLISHERS

AN IMPRINT OF ACADEMIC PRESS

A Harcourt Science and Technology Company

SAN FRANCISCO   SAN DIEGO   NEW YORK   BOSTON
LONDON   SYDNEY   TOKYO

Senior Editor   Denise E. M. Penrose
Director of Production and Manufacturing   Yonie Overton
Senior Production Editor   Cheri Palmer
Editorial Coordinator   Meghan O'Keefe
Cover Design   Ross Carron Design
Cover Image   © Dan McCoy/Rainbow/PictureQuest
Text Design, Composition, and Illustration   Windfall Software
Copyeditor   Carol Leyba
Proofreader   Jennifer McClain
Indexer   Ty Koontz
Printer   Courier Corporation

ACADEMIC PRESS
A Harcourt Science and Technology Company
525 B Street, Suite 1900, San Diego, CA 92101-4495, USA
http://www.academicpress.com

Academic Press
Harcourt Place, 32 Jamestown Road, London NW1 7BY United Kingdom
http://www.academicpress.com

Morgan Kaufmann Publishers
340 Pine Street, Sixth Floor, San Francisco, CA 94104-3205
http://www.mkp.com

Library of Congress Cataloging-in-Publication Data

Lala, Parag K., date.
  Self-checking and fault-tolerant digital design / Parag Lala.
    p.   cm.
  Includes bibliographical references and index.
  ISBN 0-12-434370-8 (acid-free paper)
    1. Electronic digital computers–Design and construction.   2. Fault-tolerant computing.
  3. Logic design.   4. Sequential machine theory.   I. Title.
  TK7888.3.L274   2001
  004.2—dc21   98-052157

This book is printed on acid-free paper.

To my brothers, Santi Priya and Debabrata,

and my sisters, Anima, Kaveri, and Rita

———————————

When a man does not understand a thing, he feels discord within himself: he seeks causes for this dissonance not in himself, as he should, but outside himself, and the result is war with something he does not understand.

—Anton Chekhov

Contents

Preface   xi

1   Fundamentals of Reliability   1

1.1   Reliability and the Failure Rate   2
1.2   Relation between Reliability and Mean-Time-Between-Failures   4
1.3   Maintainability   6
1.4   Availability   8
1.5   Series and Parallel Systems   8
1.6   Dependability   12
1.7   References   12

2   Error Detecting and Correcting Codes   15

2.1   Parity Code   16
2.2   Multiple Error Detecting Codes   17
    2.2.1 Unordered Codes for Unidirectional Error Detection   18
    2.2.2 t-unidirectional Error Detecting Codes   20
    2.2.3 Burst Unidirectional Error Detecting Code   22
2.3   Residue Codes   27
2.4   Cyclic Codes   28

2.5 Error Correcting Codes   30
    2.5.1 Hamming Code   31
    2.5.2 Hsiao Code   34
    2.5.3 Reed-Solomon Code   36

2.6 References   40

3   Self-Checking Combinational Logic Design   43

3.1 Strongly Fault-Secure Circuits   46

3.2 Strongly Code-Disjoint Circuits   47

3.3 Terminology   49

3.4 Bidirectional Error-Free Combinational Circuit Design   50

3.5 Detection of Input Fault Induced Bidirectional Errors   53

3.6 Techniques for Bidirectional Error Elimination   55
    3.6.1 Input Encoding   55
    3.6.2 Output Encoding   58

3.7 Self-Dual Parity Checking   61

3.8 Self-Checking Design Using Low-Cost Residue Code   65

3.9 Totally Self-Checking PLA Design   67

3.10 Fail-Safe Combinational Circuit Design   73

3.11 References   76

4   Self-Checking Checkers   79

4.1 The Two-Rail Checker   79

4.2 Totally Self-Checking Checkers for m-out-of-n Codes   82
    4.2.1 Pass Transistor-Based Checker Design for a Subset of m-out-of-$2m$
        Codes   95
    4.2.2 Totally Self-Checking Checker for 1-out-of-n Code   99

4.3 Totally Self-Checking Checker for Berger Code   107

4.4 Totally Self-Checking Checker for Low-Cost Residue Code   126

4.5 References   128

5   Self-Checking Sequential Circuit Design   131

5.1 Faults in State Machines   132

5.2   Self-Checking State Machine Design Techniques   134

5.3   Elimination of Bidirectional Errors   143

5.4   Synthesis of Redundant Fault-Free State Machines   145

5.5   Decomposition of Finite State Machines   150

5.6   Self-Checking Interacting State Machine Design   152

5.7   Fail-Safe State Machine Design   156

5.8   References   159

6   **Fault-Tolerant Design**   **161**

6.1   Hardware Redundancy   162

6.1.1 Static Redundancy   162
6.1.2 Dynamic Redundancy   168
6.1.3 Hybrid Redundancy   172

6.2   Information Redundancy   176

6.2.1 Fault-Tolerant State Machine Design Using Hamming Codes   176
6.2.2 Error Checking and Correction (ECC) in Memory Systems   178
6.2.3 Improvement in Reliability with ECC   179
6.2.4 Multiple Error Correction Using Orthogonal Latin Squares
Configuration   181
6.2.5 Soft Error Correction Using the Horizontal and Vertical Parity
Method   185

6.3   Time Redundancy   187

6.4   Software Redundancy   188

6.5   System-Level Fault Tolerance   189

6.5.1 Byzantine Fault Model   191
6.5.2 System-Level Fault Detection   194
6.5.3 Backward Recovery Schemes   196
6.5.4 Forward Recovery Schemes   196

6.6   References   198

Appendix: Markov Models   203

Index   205

Preface

The complexity of digital systems has increased dramatically over the last decade. This has been possible because more transistors can be packed on a VLSI chip by reducing their dimensions. The continuing development in processing technology will reduce transistor dimensions (already decreased from 5 microns in 1985 to less than 0.18 micron in 1999) even further in the near future. Smaller transistors in VLSI chips will have a detrimental effect on the reliability of a system because they are likely to show a high rate of transient errors resulting from cross-talk, noise, and the like during normal operation. Also, smaller transistors are more sensitive to the normal radiation on the earth's surface and may fail temporarily, thereby increasing the rate of transient faults in a chip. The traditional off-line testing strategy cannot guarantee the detection of transient faults; they have to be detected during normal operation of a chip. This in turn requires that VLSI circuits be designed such that they have built-in mechanisms for on-line fault detection (i.e., self-checking). An alternative approach to cope with faults in digital systems is to incorporate fault tolerance in the system. The objective of fault tolerance is either to mask, or to recover from, faults once they have been detected. Many fault-tolerant systems use self-checking circuits to detect faults before the recovery process is started.

This book is written primarily for graduate students in electrical/computer engineering; it will also be suitable as a text for final year options in undergraduate courses provided the readers have some background in switching theory and logic design. The book will also be useful for practicing engineers who are interested in highly reliable digital system design.

Chapter 1 introduces the basics of reliability theory. Common terms used in the reliability measure, such as mean-time-between-failures, availability, and dependability are defined. Also, the importance of maintainability is emphasized.

Chapter 2 addresses the main concepts of coding and then covers the construction of several error detecting and correcting codes. Error detecting codes have been extensively used in self-checking logic design, whereas error correcting codes have in general been used to enhance the reliability of computer memory systems.

Chapter 3 focuses on self-checking circuits. The basic concepts of self-checking logic are reviewed first, followed by an in-depth discussion of currently available techniques for designing self-checking combinational circuits. The last section deals with fail-safe design of combinational circuits.

Chapter 4 presents techniques for checker design. Checkers are subcircuits that are used for detecting the presence of erroneous bits at the outputs of self-checking circuits. Checker design techniques for various types of error detecting codes, which have been used for encoding the outputs of self-checking circuits, are discussed in detail.

Chapter 5 deals with various design techniques for realizing totally self-checking sequential circuits. Several such techniques for composite state machines are reviewed. A technique for designing interacting state machines is also included. The chapter concludes with a technique for fail-safe state machine design.

Chapter 6 covers all the major techniques for hardware fault tolerance in detail as well as the application of information redundancy in fault-tolerant memory system design. In addition, software and time redundancy techniques are outlined. Tolerance of arbitrary faults in a system based on Byzantine agreement is considered. Also, the PMC (Preparata, Metze, and Chien) model for fault detection at the system level and the use of backward and forward recovery techniques for fault tolerance in distributed systems are discussed.

I would like to thank my former students Drs. Fadi Busaba, Dali Tao, and Alvernon Walker; their work has been extremely helpful in the preparation of this book. I am also grateful to Professor Stephen Su of the State University of New York at Binghamton for his comments. My special thanks go to my Ph.D. thesis advisor Dr. J. I. Missen of the City University of London, now retired, who first introduced me to the subject and also helped me in many other ways for which I remain indebted.

Above all, I want to express my gratitude to my wife, Meena, and children, Nupur and Kunal, for their love and understanding.

Self-Checking
and Fault-Tolerant
Digital Design

1 | Fundamentals of Reliability

In recent years the complexity of digital systems has increased dramatically. Although semiconductor manufacturers try to ensure that their products are reliable, it is almost impossible not to have faults somewhere in a system at any given time. As a result, reliability has become a topic of major concern to both system designers and users [see references 1.1, 1.2]. A fundamental question in estimating reliability is whether a system will function in a prescribed manner in a given environment for a given period of time. This, of course, depends on many factors such as the design of the system, the parts and components used, and the environment. Hence it is natural to consider the reliability of a system as an unknown parameter, which is defined to be the probability that the given system will perform its required function under specified conditions for a specified period of time.

The reliability of a system can be increased by employing the method of worst case design, using high-quality components and imposing strict quality control procedures, during the assembly phase. However, such measures can increase the cost of a system significantly. An alternative approach to reliable system design is to incorporate "redundancy" (i.e., additional resources) into a system with the aim of masking the effects of faults. This approach does not necessitate the use of high-quality components; instead standard components can be used in a redundant and reconfigurable architecture. In view of the decreasing cost of hardware components, it is certainly less expensive to use the second approach to design reliable systems.

1.1

RELIABILITY AND THE
FAILURE RATE

Let us consider the degradation of a sample of N identical components under "stress conditions" (temperature, humidity, etc.). Let $S(t)$ be the number of surviving components, that is, the number of components still operating at time t after the beginning of the "aging experiment," and $F(t)$ the number of components that have failed up to time t. Then the probability of survival of the components, also known as the *reliability* $R(t)$, is

$$R(t) = \frac{S(t)}{N}$$

The probability of failure of the components, also known as the *unreliability* $Q(t)$, is

$$Qt = \frac{F(t)}{N}$$

Since $S(t) + F(t) = N$, we must have

$$R(t) + Qt = 1$$

The failure rate, also known as the *hazard rate* $Z(t)$, is defined to be the number of failures per unit time compared with the number of surviving components:

$$Z(t) = \frac{1}{S(t)} \frac{dF(t)}{dt} \tag{1.1}$$

Studies of electronic components show that under normal conditions the failure rate varies as indicated in Figure 1.1. There is an initial period of high failure because in any large collection of components there are usually components with defects, and these fail—that is, they do not work as intended—after they are put into operation. For this reason, the first period is called the *burn-in* period of defective components. The middle phase is the *useful life* period when the failure rate is relatively constant; in this phase failures are random in time. The final phase is the *wear-out* period, when the failure rate begins to increase rapidly with time. The curve of Figure 1.1 is often called the *bathtub* curve because of its shape.

In the useful life period the failure rate is constant, and therefore

$$Z(t) = \lambda \text{ (say)} \tag{1.2}$$

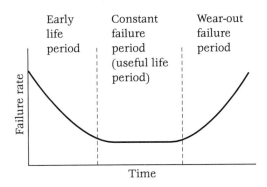

Early life period | Constant failure period (useful life period) | Wear-out failure period

1.1

Figure

Variation of failure rate with time.

With the previous nomenclature,

$$R(t) = \frac{S(t)}{N} = \frac{N - F(t)}{N} = 1 - \frac{F(t)}{N}$$

Therefore,

$$\frac{dR(t)}{dt} = -\frac{1}{N} \cdot \frac{dF(t)}{dt}$$

or

$$\frac{dF(t)}{dt} = -\frac{dR(t)}{dt} \tag{1.3}$$

Substituting Equations (1.2) and (1.3) in Equation (1.1)

$$\lambda = -\frac{N}{S(t)} \cdot \frac{dR(t)}{dt}$$

$$= -\frac{1}{R(t)} \cdot \frac{dR(t)}{dt} \quad \text{since} \quad R(t) = \frac{S(t)}{N}$$

or

$$\lambda \cdot dt = -\frac{dR(t)}{R(t)}$$

The above expression may be integrated giving

$$\lambda \int_0^t dt = -\int_1^{R(t)} \frac{dR(t)}{R(t)}$$

The limits of the integration are chosen in the following manner: $R(t)$ is 1 at $t = 0$, and at time t by definition the reliability is $R(t)$. Integrating, then,

$$\lambda |t|_0^t = |\log_c R(t)||_1^{R(t)}$$

$$\lambda t = -|\log_c R(t) - \log_c 1|$$

$$-\lambda t = \log_c R(t)$$

Therefore,

$$R(t) = \exp(-\lambda t) \tag{1.4}$$

The above relationship is generally known as the *exponential failure law*; λ is usually expressed as percentage failures per 1000 hours or as failures per hour. When the product λt is small,

$$R(t) = 1 - \lambda t \tag{1.5}$$

System failures, like component failures, can also be categorized into three periods of operation. The early system failures such as wiring errors, dry joints, faulty interconnections, and the like are normally eliminated by the manufacturer's test procedures. System failures occurring during the useful life period are entirely due to component failures.

If a system contains k types of components, each with failure rate λ_k, then the system failure rate, λ_{ov}, is

$$\lambda_{ov} = \sum_1^k N_k \lambda_k$$

where there are N_k of each type of component.

1.2   RELATION BETWEEN RELIABILITY AND MEAN-TIME-BETWEEN-FAILURES

Reliability $R(t)$ gives different values for different operating times. Since the probability that a system will perform successfully depends on the conditions under which it is operating and the time of operation, the reliability figure is not the ideal for practical use. More useful to the user is the average time a system will run between failures; this time is known as the *mean-time-between-failures*

(MTBF). The MTBF of a system is usually expressed in hours and is given by $\int_0^\infty R(t)\,dt$, that is, it is the area underneath the reliability curve $R(t)$ plotted versus t; this result is true for any failure distribution. For the exponential failure law,

$$\text{MTBF} = \int_0^\infty \exp(-\lambda t)\,dt$$

$$= -\frac{1}{\lambda}|\exp(-\lambda t)|_0^\infty = \frac{1}{\lambda} \tag{1.6}$$

In other words, the MTBF of a system is the reciprocal of the failure rate. If λ is the number of failures per hour, the MTBF is expressed in hours. If, for example, we have 4000 components with a failure rate of 0.02% per 1000 hours, the average number of failures per hour is

$$\frac{0.02}{100} \times \frac{1}{1000} \times 4000 = 8 \times 10^{-4} \text{ failures/hour}$$

The MTBF of the system is therefore equal to $1/(8 \times 10^{-4})$ or $1/8 \times 10^4 = 1250$ hours. Substituting Equation (1.6) in the reliability expression Equation (1.4) gives

$$R(t) = \exp(-\lambda t)$$

$$= \exp(-t/\text{MTBF}) \tag{1.7}$$

A graph of reliability against time is shown in Figure 1.2. As time increases the reliability decreases, and when $t = \text{MTBF}$, the reliability is only 36.8%. Thus,

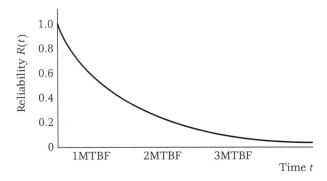

1.2   Reliability curve.

Figure

a system with an MTBF of, say, 100 hours has only a 36.8% chance of running 100 hours without failure.

By combining Equations (1.5) and (1.6), we have

$$R(t) = 1 - \lambda t$$

$$= 1 - \frac{t}{\text{MTBF}}$$

Therefore,

$$\text{MTBF} = \frac{t}{1 - R(t)}$$

Example. A first-generation computer contains 10,000 thermionic valves each with $\lambda = 0.5\%/(1000 \text{ hours})$. What is the period of 99% reliability? [1.6]

$$\text{MTBF} = \frac{t}{1 - 0.99}$$

$$t = \text{MTBF} \times 0.01$$

$$= \frac{0.01}{\lambda_{ov}} \qquad\qquad (1.8)$$

$$N = \text{number of valves} = 10{,}000$$

$$\lambda = \text{failure rate of valves} = 0.5\%/(1000 \text{ hours})$$

$$= 0.005/(1000 \text{ hours})$$

$$= 5 \times 10^{-6}/\text{hour}$$

Therefore,

$$\lambda_{ov} = N\lambda = 10^4 \times 5 \times 10^{-6} = 5 \times 10^{-2}/\text{hour}$$

From Equation (1.8),

$$t = \frac{0.01}{5 \times 10^{-2}} = \frac{10^{-2}}{5 \times 10^{-2}} = 0.2 \text{ hour} = 12 \text{ minutes}$$

This figure was often typical!

1.3   MAINTAINABILITY

When a system fails, repair action is normally carried out to restore the system to operational effectiveness. The probability that a failed system will be restored to working order within a specified time is called the *maintainability* of the system. In other words, maintainability is the probability of isolating and repairing a "fault" (see Chapter 2) in a system within a given time. There is therefore a relationship between maintainability and repair rate μ and hence with mean-time-to-repair (MTTR). MTTR and μ are always related [1.3]:

$$\mu = \frac{1}{\text{MTTR}}$$

MTTR and μ are related to maintainability $M(t)$ as follows:

$$M(t) = 1 - \exp(-\mu t) = 1 - \exp\left(-\frac{t}{\text{MTTR}}\right)$$

where t is the permissible time constraint for the maintenance action.

In order to design and manufacture a maintainable system, it is necessary to predict the MTTR for various fault conditions that could occur in the system. Such predictions are generally based on the past experiences of designers and the expertise available to handle repair work.

The system repair time consists of two separate intervals—passive repair time and active repair time [1.3]. The passive repair time is mainly determined by the time taken by service engineers to travel to the customer site. In many cases the cost of travel time exceeds the cost of the actual repair. The active repair time is directly affected by the system design and may be subdivided as follows:

1. The time between the occurrence of a failure and the system user becoming aware that it has occurred.

2. The time needed to detect a fault and isolate the replaceable component(s) responsible.

3. The time needed to replace the faulty component(s).

4. The time needed to verify that the fault has been removed and the system is fully operational.

The active repair time can be improved significantly by designing the system so that faults may be detected and quickly isolated. As more complex systems are designed, it becomes more difficult to isolate the faults. However, if adequate self-test features are incorporated into the replaceable components of a system, it becomes easier to detect and isolate faults, which facilitates repair [1.4].

1.4   AVAILABILITY

Availability of a system at time t is the probability that the system will be up (i.e., functioning according to its specification) at the instant t during its scheduled working period. Whereas reliability assumes correct operation during the time interval $(0, t)$, availability requires correct operation only for the instant t. A system can go through several repairs during the time interval $(0, t)$, and as long as the repair time is short the availability of the system will be high. The relationship between availability and repair time can be derived as follows [1.3]:

$$\text{Availability} = \frac{\text{System up-time}}{\text{System up-time} + \text{System down-time}}$$

$$= \frac{\text{System up-time}}{\text{System up-time} + (\text{Number of failures} \times \text{MTTR})}$$

$$= \frac{\text{System up-time}}{\text{System up-time} + (\text{System up-time} \times \lambda \times \text{MTTR})}$$

$$= \frac{1}{1 + (\lambda \times \text{MTTR})}$$

$$= \frac{\text{MTBF}}{\text{MTBF} + \text{MTTR}} \quad \text{since} \quad \lambda = \frac{1}{\text{MTBF}}$$

If the MTTR can be reduced, availability will increase and the system will be more economical. A system where faults are rapidly diagnosed is more desirable than a system that has a lower failure rate but where the cause of a failure takes a long time to locate, and consequently a lengthy system down-time is needed for repair.

1.5   SERIES AND PARALLEL
SYSTEMS

The reliability of a system can be derived in terms of the reliabilities or the failure rates of the subsystems used to build it. Two limiting cases of system design are frequently met in practice:

1. Systems in which each subsystem must function if the system as a whole is to function.

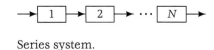

1.3

Figure

Series system.

2. Systems in which the correct operation of just one subsystem is sufficient for the system to function satisfactorily. In other words, the system consists of redundant subsystems and will fail only if all subsystems fail.

Case 1   Let us consider a system in which a failure of any subsystem would cause a system failure. This can be modeled as a series system as shown in Figure 1.3. If the subsystem failures are independent and R_i is the reliability of subsystem i, then the overall system reliability is

$$R_{\text{ov}} = \prod_{i=1}^{N} R_i$$

In the constant failure rate case where $R_i = \exp(-\lambda_i t)$

$$R_{\text{ov}} = \prod_{i=1}^{N} \exp(-\lambda_i t)$$

$$= \exp\left(\sum_{i=1}^{N} \lambda_i t\right)$$

Therefore, the failure rate of the system is just the sum of the failure rates of the subsystems.

If the N subsystems have identical failure rates $\lambda_i = \lambda$, then $R_i = R$. Hence the overall system reliability is

$$R_{\text{ov}} = \exp(-N\lambda t)$$

$$= R^N$$

and

$$\text{MTBF} = \frac{1}{N\lambda}$$

Note that the overall reliability is decreased N-fold while the MTBF is $1/N$ of that of the subsystem. For example, if each subsystem has 99% reliability after a year, a system consisting of 10 subsystems will have a reliability of 0.99^{10} or

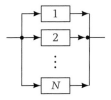

1.4

Parallel system.

Figure

about 0.9. Consequently, in a series system high reliability can be achieved only if the individual subsystems have very high reliability. ∎

Case 2 In this case system failure can occur only when all subsystems have failed. This can be modeled as a parallel system, as shown in Figure 1.4. If the failures are independent and R_i is the reliability of subsystem i, then the overall reliability of the system is

$$R_{\text{ov}} = 1 - \prod_{i=1}^{N}(1 - R_i)$$

If all the subsystems are identical, each with a constant failure rate λ, then

$$R_{\text{ov}} = 1 - (1 - R)^N$$
$$= 1 - |1 - \exp(-\lambda t)|^N$$

For example, if a system consists of 10 mutually redundant subsystems, each having only 0.75 reliability, the overall reliability of the system will be

$$R_{\text{ov}} = 1 - (1 - 0.75)^{10}$$
$$\approx 0.9999 \qquad\qquad ∎$$

In general the MTBF of a parallel system with N identical subsystems is $\sum_{j=1}^{N} 1/j$ times better than that of a single subsystem. For example, if a parallel system consists of two subsystems, then

$$R_{\text{ov}} = 1 - |1 - \exp(-\lambda t)|^2$$
$$= 2 \exp(-\lambda t) - \exp(-2\lambda t)$$

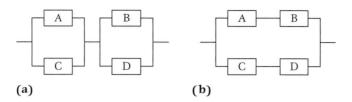

(a)   **(b)**

(a) Parallel-to-series interconnection scheme; (b) series-to-parallel interconnection scheme.

Therefore, the MTBF of the system is

$$= \int_0^\infty |2\exp(-\lambda t) - \exp(-2\lambda t)| \, dt$$

$$= 3/2\lambda$$

In practice a system normally consists of a combination of series and parallel subsystems. Figure 1.5 depicts two different interconnections of four subsystems. These systems are useful when short-circuits or open-circuits are the most commonly expected faults. The parallel-to-series network of Figure 1.5(a) is used when the primary failure mode is an open-circuit, whereas the series-to-parallel network of Figure 1.5(b) is used when the primary mode is a short-circuit [1.5].

If subsystems A and C are processors and subsystems B and D are memories, the system of Figure 1.5(a) can operate if (A, D) or (C, B) or (A, B) or (C, D) works, whereas the system of Figure 1.5(b) can operate only if either (A, B) or (C, D) works. In this situation the reliability of the parallel-to-series system is

$$R_{PS} = |1 - (1 - R_A)(1 - R_C)||1 - (1 - R_B)(1 - R_D)|$$

and the reliability of the series-to-parallel system is

$$R_{SP} = |1 - (1 - R_A R_B)(1 - R_C R_D)|$$

where R_A, R_B, R_C, and R_D are the reliabilities of subsystems A, B, C, and D, respectively. Assuming

$$R_A = R_B = R_C = R_D = R$$

R	0.7	0.8	0.9	0.95
R_{PS}	0.828	0.921	0.980	0.995
R_{SP}	0.739	0.870	0.963	0.991

1.1

Table

Comparison of R_{PS} and R_{SP}.

then

$$R_{PS} = R^4 - 4R^3 + 4R^2$$

and

$$R_{SP} = 2R^2 - R^4$$

Some indication of the effectiveness of the series-to-parallel and parallel-to-series schemes is shown by assigning a range of values to R, as in Table 1.1. The figures in the table show clearly that $R_{PS} > R_{SP}$.

1.6 DEPENDABILITY

Dependability is a framework that incorporates reliability and availability as distinct features of system specification [1.6]. Either reliability or availability can describe a system's dependability based on its characteristics. If the system is nonrepairable, then its dependability can be characterized by its reliability; on the other hand, if the system is repairable, availability can be used to characterize it.

Several software packages have been developed for analyzing the dependability of complex systems [1.7]. Most of these require reliability block diagrams, fault trees, or Markov chains as inputs. Unfortunately these forms of input models are extremely difficult to develop for multiprocessor systems, which are being increasingly used in critical applications. More research is needed for efficient evaluation of dependability in multiprocessor systems [1.8].

1.7 REFERENCES

[1.1]   Champine, G. A. "What makes a system reliable?" *Datamation*, September 1978, 195–206.

[1.2]   *IEEE Spectrum.* Special issue on Reliability (October 1981).

[1.3]   Smith, D. J. *Reliability Engineering.* Pitman, 1972.

[1.4]   Thomas, J. L. "Modular maintenance design concept." *Proc. IEEE Micro-Delcon,* 1979, 98–103.

[1.5]   McConnel, S., and D. P. Siewiorek. "Evaluation criteria." Chap. 5 of *The Theory and Practice of Reliable System Design* (Eds.: D. P. Siewiorek and R. S. Swarz). Digital Press, 1982.

[1.6]   Laprie, J. C., and A. Coste. "Dependability: A unifying concept for reliable computing." *Proc. 12th Int. Symp. Fault-Tolerant Computing,* 1982, 18–21.

[1.7]   Johnson, A. M., and M. Malek. "Survey of software tools for evaluating reliability, availability, and serviceability." *ACM Computing Surveys* 20, no. 4 (Dec. 1988): 227–269.

[1.8]   Das, C. R., J. T. Kreulen, and M. J. Thazhuthaveetil. "Dependability modeling for multiprocessors." *IEEE Computer,* October 1990, 7–19.

[1.9]   Missen, J. I. Private communication.

2

Error Detecting and Correcting Codes

Digital systems use data in the form of a group of bits for their internal operations. Since there is a possibility that during information processing or storage, data can get corrupted due to physical defects in the system, there should be some provisions in the system for detecting erroneous bits in data. It also may be necessary to correct errors in data in order to restore the system to its normal operating mode. This typically requires additional (i.e., redundant) bits to be appended to the data or information bits for error detection and/or correction. Thus, the *length*—the number of bits in encoded data, also known as a *code word*—is greater than that of the original data.

The process of appending check bits to the information bits is called *encoding;* the opposite process—extracting the original information bits from a code word—is known as *decoding.* The ratio of the number of information bits to the number of bits in code words of a particular code is known as the *code rate.* Thus, an n-bit code obtained by encoding information bits of length k has $2k$ valid code words, $(2^n - 2^k)$ invalid (i.e., non-code) words, and the code rate k/n. The set of 2^k valid code words is called a *block code.* The primary requirements of a code are as follows [2.1]:

1. It detects all likely errors.

2. It achieves the desired degree of error detection by using minimum redundancy.

3. The encoding and decoding process is fast and simple.

Codes can be classified as either *separable* or *nonseparable*. The information bits in a separable code can be separately identified from the check bits; thus, decoding is not necessary to retrieve the information bits from a code word. On the other hand, in a nonseparable code the information bits are embedded in a code word and can only be extracted by using a decoder. Because of this, nonseparable codes have not found wide application; however, there are certain applications, such as input/output encodings and state assignments in sequential circuits, in which nonseparable codes can be used efficiently (see Chapter 5). A separable code with k information bits is said to be *systematic* if all 2^k patterns of information bits occur in code words [2.2].

2.1   PARITY CODE

One widely known separable code is the parity code. The parity code is obtained by counting the number of 1s in information bits and tacking a 0 or a 1 to make the count odd or even. Thus, for information bits 0 0 1 1, in which there are two 1s, a 0 is appended to the information bits for maintaining an even number of 1 bits; a 1 is appended to make the number of 1s odd. Even though the odd and the even parity checks are mathematically equivalent, the odd parity is generally preferred because it ensures at least a single 1 in any code. The encoding of a parity code is fast and straightforward, since only a tree of EX-OR gates is required. Also, the information bits in a code word can be processed without decoding. However, a parity check can detect only odd numbers of errors. If a double bit occurs in information bits, the parity is unchanged, leaving the error undetected.

The error detection capability of parity checking may be expanded by including a parity bit for each byte of information bits [2.3]. If the information bits are more than 1 byte long, a combination of odd and even parities for the bytes can simplify the detection of an all-0s or all-1s condition. A variation of this scheme is based on partitioning the information bits into several blocks, with each bit appearing in more than one block, and computing the parity for each block. This overlapping of parity bits not only detects more than 1-bit errors, but in the case of a single erroneous bit the location of the bit is also identified.

2.2   MULTIPLE ERROR DETECTING CODES

Faults in VLSI circuits can give rise to many types of errors. Error detecting codes are classified according to the types of errors they can detect. For example, single-bit error detecting codes can handle only random errors. However, in many cases errors that occur in logic circuits and memory systems are of a multiple nature. Multiple errors belong to one of the following classes: *symmetric, unsymmetric,* and *unidirectional.*

+ Symmetric errors: Both $0 \rightarrow 1$ and $1 \rightarrow 0$ errors can occur with equal probability in a code word.

+ Unsymmetric errors: Only one type of error—$0 \rightarrow 1$ or $1 \rightarrow 0$ but not both— can occur in a code word.

+ Unidirectional errors: Both $0 \rightarrow 1$ and $1 \rightarrow 0$ errors can occur, but they do not occur simultaneously in any code word.

Many faults in VLSI circuits have been found to cause unidirectional errors [2.4–2.6]. This has led to the development of several unidirectional error detecting codes. Before reviewing such codes, let us introduce some definitions and notations.

Let X and Y be two binary k-tuples. We denote $N(X, Y)$ as the number of $1 \rightarrow 0$ crossovers from X to Y. For example, if $X = 1\,0\,1\,0\,1\,0$ and $Y = 1\,1\,0\,1\,0\,1$, then $N(X, Y) = 2$ and $N(Y, X) = 3$. The number of bits in which two distinct binary tuples differ is known as the *Hamming distance* of a code. Thus, the Hamming distance $d(X, Y)$ between X and Y is $d(X, Y) = N(X, Y) + N(Y, X) = 5$.

A word $X(x_1, x_2, \ldots, x_k)$ *covers* another word $Y(y_1, y_2, \ldots, y_k)$, written symbolically as $X \geq Y$, if $y_i = 1$ implies $x_i = 1$ for $i = 1, 2, \ldots, k$. In other words, the positions of 1 in Y are a subset of the positions of 1 in X. For example, if $X = 1\,0\,1\,0\,1\,0$ and $Y = 1\,0\,1\,0\,0\,0$, then $X \geq Y$. It should be clear that if X covers Y, then $N(Y, X) = 0$. If X does not cover Y, and Y does not cover X, then X and Y are *unordered.* A code in which no code word is covered by any other code word is known to be an *unordered code* [2.7].

2.2.1 Unordered Codes for Unidirectional Error Detection

An unordered code is capable of detecting all unidirectional errors [2.8]. This is because in such a code, a unidirectional error cannot transform one code word into another code word. Unordered codes can be separable or nonseparable. For example, both *m-out-of-n* code and *Berger code* are unordered, but the former is nonseparable and the latter is separable. Both codes detect single and unidirectional multiple errors.

m-*out-of*-n *Codes*

In an *m-out-of-n* code, all valid code words have exactly m 1s and $(n - m)$ 0s. The total number of valid code words is $n!/(n - m)!m!$. For example, the 2-out-of-4 code has six valid code words: $0\,0\,1\,1, 0\,1\,0\,1, 1\,0\,0\,1, 0\,1\,1\,0, 1\,0\,1\,0, 1\,1\,0\,0$. An *m*-out-of-*n* code with $m = k$ and $n = 2k$, where k is an integer, is popularly known as a *k-out-of-2k* code. A special case of *k*-out-of-2*k* code consisting of only 2^k code words out of the possible $2k!/k!k!$ code words is known as a *k-pair two-rail code*. Each code word in a *k*-pair two-rail code has k information bits and k check bits, which are bit-by-bit complements of the information bits. The 2-pair two-rail code consists of the following code words: $0\,0\,1\,1, 1\,0\,0\,1, 0\,1\,1\,0, 1\,1\,0\,0$.

If $m = \lfloor n/2 \rfloor$ in an *m-out-of-n* code, then the code is optimal. In other words, there is no other unordered code except $\lfloor n/2 \rfloor$-*out-of-n* code that has more code words of length n [2.9]. A particular subset of *m*-out-of-*n* code is 1-out-of-*n* code, in which exactly 1 bit of an *n*-bit code word is 1 and the remaining bits are all 0s.

Berger Code

A Berger code of length n has k infomation bits and c check bits where $c = \lceil \log_2(k + 1) \rceil$ and $n = k + c$ [2.10]. It is the least redundant unordered code for detecting single and unidirectional multibit errors. A code word is constructed by forming a binary number corresponding to the number of 1s in the information bits, and appending the bit-by-bit complement of the binary number as check bits to the information bits. For example, if $k = 0\,1\,0\,1\,0\,0\,0$, $c = \lceil \log_2(7 + 1) \rceil = 2$ and hence the Berger code must have a length of $10(= 7 + 3)$, c check bits are derived as follows:

k	$I0$	c_1	c_2
0 1 0 1 1 0 0 1	4	0 0	1 1
0 1 0 1 1 0 1 1	3	1 1	0 0
1 1 1 1 0 0 1 1	2	1 0	0 1
1 1 0 1 1 1 1 1	1	0 1	1 0
1 1 1 1 1 1 1 1	0	0 0	1 1

2.1

Table

A subset of modified Berger code.

Number of 1s in information bits $k = 2$

Binary equivalent of $2 = 0\,1\,0$

The bit-by-bit complement of $0\,1\,0$ is $1\,0\,1$, which are the c check bits. Thus,

$$n = \underbrace{0\,1\,0\,1\,0\,0\,0}_{k}\ \underbrace{1\,0\,1}_{c}$$

It should be clear from the above discussion that the c check bits may be the binary number representing the number of 0s in k information bits. Thus, the check bits for Berger codes can be generated by using two different schemes. The scheme that uses the bit-by-bit complement of the binary representation of the number of 1s in the information bits is known as the the *B1 encoding scheme.* The other scheme, which uses the binary representation of the number of 0s in the information bits as check bits, is identified as the *B0 encoding scheme.*

If the number of information bits in a Berger code is $I = 2^c - 1$, $c \geq 1$, then it is called a *maximal length Berger code;* otherwise it is known as the *non-maximal length Berger code.* For example, the Berger code $0\,1\,0\,1\,0\,0\,0\,1\,0\,1$ is maximal length because $c = 3$ and $k = 7 = (2^c - 1)$, whereas $1\,1\,0\,1\,0\,0\,0\,1\,1$ is non-maximal length because $c = 3$ and $k = 6 \neq (2^c - 1)$.

A modified form of Berger code that requires fewer check bits than in the conventional Berger code has been proposed by Dong [2.11]. In the modified code, all unidirectional errors of length t in the information bits k can be detected if the check bits define a number $J = I0 \bmod(t + 1)$, where $I0$ is the number of 0s in the information bits. The number of check bits c_1 in the modified code is equal to $\lceil \log_2(t + 1) \rceil$ and represents the binary equivalent of J. The check bits may also be encoded using additional check bits c_2 to form another code, for example, two-rail code for error detection. Table 2.1 shows a subset of modified Berger code for $k = 8$, $t = 3$, $c_1 = 2$, and $c_2 = 2$.

2.2.2   *t*-unidirectional Error Detecting Codes

In general the code construction for detecting unidirectional errors has focused on detecting errors of multiplicity t (i.e., t-unidirectional errors) rather than all possible unidirectional errors. In this section we discuss methods for designing t-unidirectional error detecting codes proposed by Borden, and Bose and Lin.

Borden Code

Borden [2.12] has proposed a t-unidirectional error detecting code in which code words of length n have weight w that is congruent to $\lfloor n/2 \rfloor \bmod(t + 1)$. Thus, the number of code words denoted by $C(n, t)$ is

$$C(n, t) = \sum_{w = \lfloor n/2 \rfloor \bmod(t+1)} \binom{n}{w}$$

To illustrate let us assume $n = 12$ and $t = 3$. The number of code words in the $C(12, 3)$ are

$$\binom{12}{2} + \binom{12}{6} + \binom{12}{10} = 1056$$

If t is assumed to be 1, the Borden code, like the conventional parity code, can guarantee the detection of all single-bit errors. When $t = n/2$, a Borden code becomes a constant weight code (e.g., m-out-of-$2m$ code) and detects all single and unidirectional errors. Although the Borden code is nonseparable, it has the highest possible information rate k/n for any information block of length k; thus, the code is optimal.

Bose-Lin Codes

Bose and Lin [2.13] have proposed codes capable of detecting unidirectional errors of multiplicity 2, 3, and 6 using 2, 3, and 4 check bits, respectively; these codes are optimal. The check bits for 2-unidirectional and 3-unidirectional error detecting codes are $k_0 \bmod 4$ and $k_0 \bmod 8$, respectively, where k_0 is the number of 0s in the information bits. For 6-unidirectional error detecting code, the check bits are $(k_0 \bmod 8) + 4$. Thus, the check bits required to detect 2-, 3-, and 6-bit unidirectional errors in the information vector

1 0 1 1 0 1 0 1 1 1 0 0 0 1 1 1

are 1 0(= 2), 1 1 0(= 6), and 1 0 1 0(= 10), respectively.

Bose and Lin [2.13] have also shown that the multiplicity of errors that can be detected by a unidirectional error detecting code depends on how check bits are derived. They have proposed two methods for constructing code for cases in which five or more check bits are needed for encoding.

Method 1

1. The check bits c needed to detect 2- and 3-unidirectional errors are k_0 mod 2^2 and k_0 mod 2^3, respectively, where k_0 is the number of 0s in the information bits.

2. If four or more check bits are needed, they are partitioned into two blocks with the two most significant bits in the first block and the remaining $(c - 2)$ bits in the second block. First, k_0 mod $2^{c-1} = (q_{c-2}, \ldots, q_1, q_0)$ is derived. Then the actual check bits

$$p_{c-1}, \ldots, p_1, p_0$$

are derived, where $p_{c-1} = q_{c-2}$, $p_{c-2} = \overline{q}_{c-2}$, and $p_j = q_j$ for any other j.

The maximum number of unidirectional errors that can be detected by this coding method is $2^{c-2} + c - 2$.

To illustrate let us derive four check bits for information bits

$$1\,0\,1\,1\,1\,1\,1\,1\,1\,0\,0\,0\,0\,0\,1\,1$$

Thus,

$$k_0 \text{ mod } 2^{c-1} = 6 \text{ mod } 2^{4-1} = 1\,1\,0 = q_2 q_1 q_0$$

Therefore, the check bits are $p_3 p_2 p_1 p_0 = 1\,0\,1\,0$. ∎

Method 2   This method is used if five or more check bits are needed. First, $k_0 \text{ mod}(3.2^{c-3}) = (p_{c-2}, p_{c-3}, \ldots, p_1, p_0)$ is derived. Then, bits $p_{c-2} p_{c-3} p_{c-4}$, which can be one of the patterns in the set $\{0\,0\,0, 0\,0\,1, 0\,1\,0, 0\,1\,1, 1\,0\,0, 1\,0\,1\}$, are mapped into a 2-out-of-4 code. A possible mapping of the $p_{c-2} p_{c-3} p_{c-4}$ bits to six 2-out-of-4 code words is shown below:

$$0\,0\,0 \rightarrow 1\,0\,1\,0, 0\,0\,1 \rightarrow 0\,1\,0\,1, 0\,1\,0 \rightarrow 1\,0\,0\,1$$

$$0\,1\,1 \rightarrow 0\,1\,1\,0, 1\,0\,0 \rightarrow 0\,0\,1\,1, 1\,0\,1 \rightarrow 1\,1\,0\,0$$

Such a code word is assigned as the most significant bits of the check bits. The remaining $(c - 4)$ check bits take the unchanged values of $p_{c-5} \ldots p_0$. The code generated by this method can detect a unidirectional error of up to $(5 \cdot 2^{c-4} + c - 4)$ bits.

To illustrate let us assume that five check bits are to be derived for information bits

0 0 0 1 1 1 1 1 1 0 0 0 0 0 0 1

Thus, $k_0 \bmod(3 \cdot 2^{c-3}) = 9 \bmod(3 \cdot 2^{5-3}) = 1\,0\,0\,1 = (p_3, p_2, p_1, p_0)$. Bits (p_3, p_2, p_1) are mapped into the code word 0 0 1 1. Therefore, the check bits are 0 0 1 1 1.

■

2.2.3   Burst Unidirectional Error Detecting Code

A burst error of length b in k information bits affects a cluster of b adjacent bits. Such errors may be random or unidirectional. Berger [2.14] proposed a coding scheme that allows detection of random bursts of length b and an arbitrary number of unidirectional errors. To illustrate the coding scheme let

$$x_{k-1}, x_{k-2}, \ldots, x_1, x_0$$

be the information bits of length k. The number of check bits (r) needed to encode the information bits is

$$\cong b + \log_2(k/b)$$

These bits are used to compute the decimal equivalent of the check bits

$$= \sum_{i=0}^{k-1} x_i 2^{i \bmod b}$$

If $c_{r-1} \cdot c_{r-2}, \ldots, c_1, c_0$ are the check bits, then the code word corresponding to the information bits $x_{k-1}, x_{k-2}, \ldots, x_1, x_0$ is

$$x_{k-1}, x_{k-2}, \ldots, x_1, x_0, c_{r-1} \cdot c_{r-2}, \ldots, c_1, c_0$$

To illustrate the encoding process let

$$x_7, x_6, x_5, x_4, x_3, x_2, x_1, x_0 = 0\,1\,0\,0\,1\,0\,0\,1$$

be the information bits and $b = 4$. The number of check bits needed is

$$r = 4 + \log_2(8/4) = 5$$

The decimal equivalent of the check bits

$$= \sum_{i=0}^{7} x_i 2^i \bmod 4$$

$$= x_0 \cdot 2^0 + x_1 \cdot 2^1 + x_2 \cdot 2^2 + x_3 \cdot 2^3 + x_4 \cdot 2^0 + x_5 \cdot 2^1 + x_6 \cdot 2^2 + x_7 \cdot 2^3$$

$$= 1 \cdot 2^0 + 0 \cdot 2^1 + 0 \cdot 2^2 + 1 \cdot 2^3 + 0 \cdot 2^0 + 0 \cdot 2^1 + 1 \cdot 2^2 + 0 \cdot 2^3$$

$$= 13$$

Therefore, the check bits are

$$c_4, c_3, c_2, c_1, c_0 = 0\,1\,1\,0\,1$$

and hence the code word is

$$0\,1\,0\,0\,1\,0\,0\,1\,0\,1\,1\,0\,1$$

Any random burst error of length 4 or less in the code word will be detected. To illustrate let us assume that the code word changes to

$$0\,1\,0\,0\,0\,1\,1\,0\,0\,1\,1\,0\,1$$

The resulting check bits (i.e., $0\,1\,0\,1\,0$) are different from the expected check bits (i.e., $0\,1\,1\,0\,1$), indicating the presence of the burst error.

Berger [2.14] has demonstrated that a code with check bits derived as shown above is capable of detecting all unidirectional errors. Also, if the length of a burst error is assumed to be 1, the check bits correspond to the binary representation of the number of 1s in the information bits. Thus, the resulting code is the Berger code discussed in Section 2.2.1 above.

Bose [2.15] has proposed a code capable of detecting unidirectional burst errors of length up to 2^{r-1} bits using r check bits. Let us assume $x_{k-1}, x_{k-2}, \ldots, x_1, x_0$ to be the k information bits, and $c_{r-1}, c_{r-2}, \ldots, c_1, c_0$ to be the r check bits. The check bits represent the binary equivalent of the decimal number

(number of 0s in information bits) mod 2^r

The bits of a code word are arranged as follows:

$$c_0\, c_1 \ldots c_{r-3}\, c_{r-2}\, x_{k-1}\, x_{k-2} \ldots x_{k-2}^{r-1}\, c_{r-1}\, x_{k-2}^{r-1} - 1 \ldots x_1\, x_0$$

To illustrate let us assume information bits to be

$$x_7, x_6, x_5, x_4, x_3, x_2, x_1, x_0 = 1\,0\,1\,0\,0\,0\,0\,0$$

There are six 0s in the information bits. Assuming c_2, c_1, and c_0 are the check bits, we get

$$c_2 \, c_1 \, c_0 = 1 \, 1 \, 0 \, (= 6 \bmod 2^3)$$

Therefore, the code word is

$$c_0 \, c_1 \, x_7 \, x_6 \, x_5 \, x_4 \, c_2 \, x_3 \, x_2 \, x_1 \, x_0 = 0 \, 1 \, 1 \, 0 \, 1 \, 0 \, 1 \, 0 \, 0 \, 0 \, 0$$

Blaum [2.16] has proposed a unidirectional burst error detecting code that can detect burst errors of length b, where $b \geq 2^{r-1}$ and r is the number of check bits. The Blaum code is more efficient than t-unidirectional error detecting codes for $r \geq 3$ and $k \geq 2^r$. It is equivalent to the burst error detecting code proposed by Bose for $r = 2$ or 3; however, it is more efficient than the Bose code for $r \geq 4$. For example, if $r = 4$ the Blaum code can detect a burst error of up to 9 bits whereas the Bose code can detect a burst of 8 bits. The improvement is more significant for higher values of r. With $r = 5$ and 6, the Blaum code can detect burst errors of length 19 and 41, respectively, while the Bose code with the same number of check bits can detect burst errors of up to 16 and 32 bits, respectively.

Assuming $x_{k-1}, x_{k-2}, \ldots, x_1, x_0$ to be the k information bits, and $c_{r-1}, c_{r-2} \ldots c_1, c_0$ to be the r check bits, the code words in the Blaum code are arranged as follows:

$$c_0 \, x_{k-1} \cdots x_{(r-1)(b-1)} \, c_1 \, x_{(r-1)(b-1)-1} \cdots x_{(r-2)(b-1)} \, c_2 \cdots c_{r-3} \, x_{3(b-1)-1} \cdots$$

$$x_{2(b-1)+1} \, x_{2(b-1)} \, c_{(r-2)} \, x_{2(b-1)-1} \cdots x_b \, x_{(b-1)} \, c_{(r-1)} \, x_{(b-2)} \cdots x_1 \, x_0 \qquad (2.1)$$

where b is the length of the burst error that can be detected.

The check bits are derived based on the *weight,* that is, the number of 1s in the information bits. The 2^r combinations of r check bits are partitioned into $(r + 1)$ blocks. A block contains $\binom{r}{r-m}$ patterns of 2^r combinations, where $m = 0, 1, \ldots, r$; all patterns in block m have $(r - m)$ 1s. Information vectors are assigned check bit patterns in sequence, starting with the vector with weight 0 (i.e., an all-0 vector). Check bit patterns from a block are assigned, in order of their increased value, to the information vectors. To illustrate let us assume $r = 4$. Thus, there are 16 check bit patterns. These are partitioned into five blocks: block 0, block 1, block 2, block 3, and block 4, which contain 1, 4, 6, 4, and 1 patterns, respectively. The patterns in each block are shown below:

Block	Check bit patterns
0	1 1 1 1
1	0 1 1 1
	1 0 1 1
	1 1 0 1
	1 1 1 0
2	0 0 1 1
	0 1 0 1
	0 1 1 0
	1 0 0 1
	1 0 1 0
	1 1 0 0
3	0 0 0 1
	0 0 1 0
	0 1 0 0
	1 0 0 0
4	0 0 0 0

To encode a k-bit information vector, let $i \varepsilon Z$ where $Z = \{0, 1, 2, \ldots, (2^r - 1)\}$. Also, let f be the binary pattern corresponding to an element in Z. Next, $c(i, f; r)$ is defined as follows:

$$c(i, f; r) = \min\{j : 1 \leq j \leq 2^r, f(i) \leq f(i + j) \bmod 2^r),$$
$$((\text{weight } f(i + j) \bmod 2^r) - \text{weight } f(i)) \leq 1\}$$

where "$\leq$" indicates $f(i + j) \bmod 2^r$ covers $f(i)$.

Table 2.2 shows the five blocks corresponding to $r = 4$; the check bit patterns are numbered consecutively with $i = 0, 1, \ldots,$ etc. The values of $c(i, f; r)$ are written in the last column.

Let us illustrate the calculation of $c(i, f; r)$ for $i = 7$; only $j = 10$ satisfies the condition

$$c(7, 0\,1\,1\,0; 4) = \min\{j = 10 : 1 \leq j \leq 16, f(7) \leq f(7 + 10) \bmod 16),$$
$$((\text{weight } f(7 + 10) \bmod 16) - \text{weight } f(7)) \leq 1\}$$
$$= 10$$

Block	i	$f(i)$	$c(i, f; 4)$
0	0	1 1 1 1	16
1	1	0 1 1 1	15
	2	1 0 1 1	14
	3	1 1 0 1	13
	4	1 1 1 0	12
2	5	0 0 1 1	12
	6	0 1 0 1	11
	7	0 1 1 0	10
	8	1 0 0 1	10
	9	1 0 1 0	9
	10	1 1 0 0	9
3	11	0 0 0 1	10
	12	0 0 1 0	9
	13	0 1 0 0	9
	14	1 0 0 0	10
4	15	0 0 0 0	12

2.2

Table

Check bits for detecting burst errors of length up to 9 bits.

It can be seen from the last column of Table 2.2 that the minimum value of $c(i, f; 4)$, denoted by $c(f; 4)$, is 9. Thus, a burst error of length up to $9(= b)$ bits can be detected in a Blaum code constructed using four check bits.

The check bits for Blaum codes can be derived from the value of the following function and the table corresponding to $c(i, f; r)$:

$$= f(\text{weight } (x_{k-1}, x_{k-2}, \ldots, x_1, x_0) \bmod 2^r)$$

Let us illustrate the derivation of four check bits, $r = 4$, for the following information vector of length 16.

x_{15}	x_{14}	x_{13}	x_{12}	x_{11}	x_{10}	x_9	x_8	x_7	x_6	x_5	x_4	x_3	x_2	x_1	x_0
1	0	0	0	1	0	1	0	0	0	1	1	1	0	0	1

Since the information vector has a weight of 7, the value of the above function can be derived from Table 2.2:

$$= f(7 \bmod 16) = f(7) = 0 1 1 0$$

The information bits and the check bits are distributed according to Equation (2.1) to form the code word.

2.3   RESIDUE CODES

Residue codes are separable codes and in general have been used to detect errors in the results produced by arithmetic operations [2.1]. A *residue* is simply defined as the remainder after a division. The residue representation of an integer N may be obtained from

$$N = Im + r$$

where m is a check base and I is an integer, so that $0 \leq r \leq m$. The quantity r is called the residue $N \mod m$ or

$$r = N \mod m$$

In residue codes the information bits define a number $N| = \sum_{i=0}^{n} a_i(2^i)$; $a_i \varepsilon \{0, 1\}|$ and the check bits define a number $C(= N \mod m)$. The number of check bits is given by $\lceil \log_2 m \rceil$.

For example, if the information bits are $1\ 1\ 1\ 0$ and $m = 2$, then $\log_2 \lceil 2 \rceil = 2$ check bits are required. Their values are given by

$$C = (14) \mod 2$$

$$= 2, \text{i.e., } 10 \text{ in binary}$$

One class of residue codes, known as *low-cost residue codes*, has been found to be very effective in on-line detection of single-bit errors in logic circuits (see Chapter 3). A low-cost residue code is obtained by making the check base m to have the form

$$m = 2^r - 1 \quad r \geq 2$$

The check bits for such codes are of length r and are obtained by performing a $\mod(2^r - 1)$ addition of the information bits. This is usually done by dividing the number of information bits into a number of groups, each containing r bits. A $\mod(2^r - 1)$ addition of the r-bit groups results in the check bits.

For example, let us compute the check bits for information bits $1\ 0\ 0\ 1\ 1\ 1\ 1\ 0$, assuming $r = 2$. The information bits must then be divided into four 2-bit groups. The groups are added together in modulo $2(= 2^2 - 1)$ binary adders with end-around carry. Thus, the check bits for the given information bits are

$$10 + 01 + 11 + 10$$
$$= 11 + 10$$
$$= 10$$

If the number of information bits is not divisible by r, one of the adders gets fewer than $2r$ inputs; the missing bits are assumed to be 0s. For example, the check bits for information bits 10101100101 are generated by setting the missing bits equal to 0 as shown below:

$$-1 + 01 + 01 + 10 + 01 + 01$$
$$= 10 + 11 + 10$$
$$= 01$$

2.4 CYCLIC CODES

Cyclic codes are a class of linear codes with the property that any code word shifted cyclically (an end-around carry) will also result in a code word. For example, if $(c_{n-1}, c_{n-2}, \ldots, c_1, c_0)$ is a code word, then $(c_{n-2}, \ldots, c_1, c_0, c_{n-1})$ and $(c_{n-3}, \ldots, c_0, c_{n-1}, c_{n-2})$ are also code words. An (n, k) cyclic code—that is, a cyclic code with k information bits and $(n-k)$ check bits—must be a multiple of a generator polynomial $G(X)$ of degree $(n-k)$.

A systematic form of a cyclic code $c(X)$ corresponding to a data polynomial $D(X)$ can be obtained by dividing $x^{n-k}D(X)$ by $G(X)$. In other words,

$$c(X) = \frac{x^{n-k}D(X)}{G(X)} = Q(X) + \frac{R(X)}{G(X)}$$

where $Q(X)$ is the quotient and $R(X)$ is the remainder. This expression can be rewritten as

$$c(X) = x^{n-k}D(X) = Q(X)G(X) + R(X)$$

Since in modulo-2 arithmetic addition and subtraction are the same, then

$$c(X) = x^{n-k}D(X) + R(X) = G(X)Q(X)$$

Because $c(X)$ is a multiple of $G(X)$, it is a code polynomial. Thus, a data polynomial can be encoded in a systematic form by shifting it $(n-k)$ bits to the left, dividing by the generator polynomial, and using the resulting remainder bits as check bits.

To illustrate the construction of the systematic cyclic code, let us assume a code for which $n = 11$, $k = 8$, and that uses the generator polynomial

$$G(X) = x^3 + x + 1$$

If the data bits are $1\,0\,1\,0\,1\,1\,0\,0$, the corresponding data polynomial is

$$D(X) = x^7 + x^5 + x^3 + x^2$$

Therefore,

$$x^{n-k}D(X) = x^3 D(X) = x^{10} + x^8 + x^6 + x^5$$

The remainder can be obtained by dividing $x^{n-k}D(X)$ by $G(X)$,

$$R(X) = x + 1$$

Thus, the code polynomial $c(X)$ can be formed by adding $R(X)$ to $x^3 D(X)$

$$c(X) = (x^{10} + x^8 + x^6 + x^5) + (x + 1)$$
$$= \underbrace{1\,0\,1\,0\,1\,1\,0\,0}_{\text{data bits}} \quad \underbrace{0\,1\,1}_{\text{check bits}}$$

The generator matrix of the (11, 8) systematic cyclic code can be derived by dividing $x^k \cdot x^r$ by $G(X)$, where $k(7, \ldots, 0)$ and $r(= 3)$ are the positions of the data bits and the degree of the generator polynomial, respectively. The first row of the generator matrix corresponds to $k = 7$; the remainder resulting from dividing $x^7 \cdot x^3$ by $G(X)$ results in $R(X) = x + 1$. Thus, the code word corresponding to data x^7 is $x^7 \cdot x^3 + x + 1$. The other rows of the generator matrix are formed in a similar manner corresponding to data $x^6, x^5, x^4, x^3, x^2, x$, and 1 in that order. The resulting generator matrix is

$$
\begin{vmatrix}
1 & 0 & 0 & 0 & 0 & 0 & 0 & 0 & 0 & 1 & 1 \\
0 & 1 & 0 & 0 & 0 & 0 & 0 & 0 & 1 & 0 & 0 \\
0 & 0 & 1 & 0 & 0 & 0 & 0 & 0 & 0 & 1 & 0 \\
0 & 0 & 0 & 1 & 0 & 0 & 0 & 0 & 0 & 0 & 1 \\
0 & 0 & 0 & 0 & 1 & 0 & 0 & 0 & 1 & 0 & 1 \\
0 & 0 & 0 & 0 & 0 & 1 & 0 & 0 & 1 & 1 & 1 \\
0 & 0 & 0 & 0 & 0 & 0 & 1 & 0 & 1 & 1 & 0 \\
0 & 0 & 0 & 0 & 0 & 0 & 0 & 1 & 0 & 1 & 1
\end{vmatrix}
$$

As discussed previously, the basic idea of coding is to add check bits to the information bits such that if an error occurs in the data bits, it can be detected. In certain applications, such as in computer memory systems, the on-line correction of a detected error becomes indispensable. An error correcting code not only has the capability to detect bit error(s), but can also identify the location of the erroneous bits.

The error detecting and correcting capability of a code can be defined in terms of the Hamming distance of a code. The relationship between the Hamming distance of a code and its error detecting and correcting capabilities can be defined as

$$d = C + D + 1 \quad \text{with} \quad D \geq C$$

where

D = number of bit errors that can be detected

C = number of bit errors that can be corrected

The relationship between C and D for values of d up to 5 is shown in Table 2.3. In codes with $d = 1$, a single error in a code word can result in another code word, thereby preventing the detection of single errors. In codes with $d = 2$, a single-bit error will convert a code word into a non-code word; however, two or more errors may convert it into another valid code word. Thus, codes with $d = 2$ are single error detecting codes. In codes with $d = 3$, all code words differ in at least three bit positions. A single error in a code word will result in a non-code word (i.e., a binary tuple that will have a Hamming distance of 1 from one code word, and a distance of 2 or more from all others).

For example, Figure 2.1 shows two code words $C_1 (= 1\,1\,0\,1\,0\,1\,0\,0)$ and $C_2 (= 0\,0\,1\,1\,1\,1\,0\,0)$ for a code with $d = 4$. If code word C_1 changes to non-code word $NC_1 (= 1\,1\,0\,1\,1\,1\,0\,0)$ due to a single-bit error, the erroneous bit can be detected and corrected $(C = 1)$. Similarly, if a single-bit error changes C_2 to non-code word $NC_2 (= 0\,0\,0\,1\,1\,1\,0\,0)$, the error can be detected and corrected $(C = 1)$. However, C_1 and C_2 can be converted into the non-code word $NC_3 (= 0\,1\,0\,1\,1\,1\,0\,0)$ if the fourth and the eighth bits of C_1 are erroneous or the sixth and the seventh bits of C_2 are erroneous. Since either of the double-bit errors could have produced NC_3, it is possible only to detect a double-bit error $(D = 2)$ but not to uniquely identify and correct the errors $(C = 0)$.

The key to error correction is that it should be possible to identify erroneous bits. Two error correcting codes that are extensively used in computer memory

d	D	C
1	0	0
2	1	0
3	1	1
	2	0
4	2	1
	3	0
5	4	0
	3	1
	2	2

2.3

Table

Relation between Hamming distance and error detecting/correcting capability.

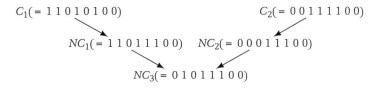

$C_1(= 1\ 1\ 0\ 1\ 0\ 1\ 0\ 0)$ $C_2(= 0\ 0\ 1\ 1\ 1\ 1\ 0\ 0)$

$NC_1(= 1\ 1\ 0\ 1\ 1\ 1\ 0\ 0)$ $NC_2(= 0\ 0\ 0\ 1\ 1\ 1\ 0\ 0)$

$NC_3(= 0\ 1\ 0\ 1\ 1\ 1\ 0\ 0)$

2.1

Figure

Two code words (with $d = 4$) and some non-code words.

systems—*Hamming code* and *Hsiao code*—use multiple parity bits to uniquely determine which bits in a code word are erroneous as well as to define the error-free condition. The structure of both of these codes is such that the linear sum of two code words is always another code word; that is, these are *linear codes*.

2.5.1   Hamming Code

A Hamming code can be described by its *parity check matrix P* of n columns, each corresponding to one of the n bits of the encoded word, and r rows, each corresponding to one of the parity check bits. The elements of the matrix are 0s and 1s; the position of the 1s in the ith row indicate which bit positions are involved in the parity check. Similarly, the position of 1s in the jth column indicate the parity pattern corresponding to the jth bit.

Single-error correcting codes have a distance of 3. The minimum number of parity check bits required for single-error correction is determined from the relationship (known as the *Hamming relationship*):

$$2^c \geq k + c + 1$$

where k = number of data bits; c = number of check bits.

The construction of a Hamming code with $k = 4$ will now be explained. It can be seen from the Hamming relationship that if $k = 4$, $c = 3$. Thus, three check bits have to be appended to the four data bits in order for the Hamming code to be single-error correcting. The bit positions of the code are labeled with numbers 1 through 7:

Bit positions	7	6	5	4	3	2	1
Bit names	b_3	b_2	b_1	c_2	b_0	c_1	c_0

The bit positions corresponding to powers of 2 are used as check bits c_0, c_1, and c_2, respectively. The other bit positions correspond to the data bits b_0 to b_3.

The parity check matrix for the Hamming code with $k = 4$ and $c = 3$ is

$$P = \begin{matrix} b_3 & b_2 & b_1 & c_2 & b_0 & c_1 & c_0 & \\ \begin{vmatrix} 1 & 1 & 1 & 1 & 0 & 0 & 0 \\ 1 & 1 & 0 & 0 & 1 & 1 & 0 \\ 1 & 0 & 1 & 0 & 1 & 0 & 1 \end{vmatrix} & \begin{matrix} e_2 \\ e_1 \\ e_0 \end{matrix} \end{matrix}$$

It can be seen from the parity check matrix that

$$c_2 = b_3 \oplus b_2 \oplus b_1$$
$$c_1 = b_3 \oplus b_2 \oplus b_0$$
$$c_0 = b_3 \oplus b_1 \oplus b_0$$

For example, if $b_3 b_2 b_1 b_0 = 0\,1\,0\,1$, then $c_2 = 1$, $c_1 = 0$, and $c_0 = 1$. Thus, the corresponding Hamming code is

Bit positions	7	6	5	4	3	2	1
	0	1	0	1	1	0	1

Suppose that the encoded word is stored in the memory and on a read operation bit 3 changes from 1 to 0. To determine whether the word is correct or not, the check bits for the word are regenerated and compared with the check bits generated before storing the word in the memory. The new check bits are

$$c_2' = 1 \text{ (for even parity)} \quad c_1' = 1 \text{ (for even parity)}$$

and

$$c_0' = 0 \text{ (for even parity)}$$

The *error address* (i.e., the location of the bit in error) is generated from the following equations:

$$e_2 = c_2 \oplus c_2', \quad e_1 = c_1 \oplus c_1', \quad \text{and} \quad e_0 = c_0 \oplus c_0'$$

Hence,

$$e_2 = 1 \oplus 1 = 0, \quad e_1 = 0 \oplus 1 = 1, \quad \text{and} \quad e_0 = 1 \oplus 0 = 1$$

Therefore, the error address is 0 1 1 (i.e., 3_{10}) and bit position 3 must be inverted. If the error address is all 0s, then no bit error has occurred in the stored word.

Hamming codes with distance 3 can detect two-bit errors but can only be used to correct single-bit errors. For example, if bits 3 and 5 are erroneous in the previous example,

Bit positions	7	6	5	4	3	2	1
	0	1	1	1	0	0	1

then $c_2' = 0$, $c_1' = 1$, and $c_0' = 1$. Hence the error address is $e_2 = 1(1 \oplus 0)$, $e_1 = 1(0 \oplus 1)$, and $e_0 = 0(1 \oplus 1)$, which points to bit position 1 1 0 ($= 6_{10}$). But bit position 6 is not erroneous! Thus, any attempt to execute double-bit error correction with distance-3 Hamming code will result in incorrect correction.

The single-error correcting Hamming code can be made into a distance-4 code with the addition of another parity check bit at bit position 8.

Bit positions	8	7	6	5	4	3	2	1
Bit names	c_3	b_3	b_2	b_1	c_2	b_0	c_1	c_0

This bit checks for even parity over the entire 8-bit word. When the overall parity check of the encoded word is correct and the error address is zero, there is no bit error. If the overall parity is wrong and the error address has a nonzero value, there is a single, correctable bit error. If the overall parity of the encoded word is correct, but the error address is not zero, there is a noncorrectable double-bit error in the word.

Table 2.4 shows the increase in word length with Hamming single-error correcting/double-error detecting codes. It can be seen from the table that the percentage increase in word length decreases with the increase of data word length. In other words, the use of error correction becomes more effective as the number of bits per word becomes larger. For example, for an 8-bit data

Data word length	Check bits	Increase
8	5	62.5%
16	6	37.5%
32	7	21.9%
48	7	14.6%
64	8	12.5%

2.4

Table

Increase in word length with single-error correction and double-error detection.

word, 5 check bits—62.5% of the data word—are required, but for a 64-bit data word, 8 check bits—12.5% of the data word—are required. The check bits are treated in exactly the same way as data bits, in that a single-bit error in either category will be detected and corrected.

2.5.2 Hsiao Code

As discussed in the previous section, a single-error correcting and double-error detecting code has a Hamming distance of at least 4. This is achieved by modifying the parity check matrix of a single-error correcting code with a row of all 1s. Hsiao code is different from Hamming code and is based on the property that the linear sum of two r-tuples with an odd number of 1s results in an r-tuple with an even number of 1s [2.18].

The parity check matrix for the Hsiao code is constructed as follows:

- ✦ Assuming r check bits $(c_{r-1}, c_{r-2}, \ldots, c_1, c_0)$, an r-bit column with a single 1 is assigned to check bit c_i; the bit position i (from top) in that column is 1; all other bit positions are at 0.

- ✦ If the length of the information bits is k and $\binom{r}{3} > k$, k columns out of $\binom{r}{3}$ combinations are selected such that each of these columns has three 1s. If $\binom{r}{3} < k$, all $\binom{r}{3}$ columns are selected; the remaining columns are selected first from among $\binom{r}{5}$ columns having five 1s, then from among $\binom{r}{7}$ columns having seven 1s, and so on. This process is continued until all k columns in the parity check matrix have been specified.

Let us illustrate the derivation of the parity check matrix for the Hsiao code for $k = 8$. The number of check bits needed for single-bit error correction and double-bit error detection is $r = 5$. Each check bit is assigned a 5-bit column as shown below:

c_4	c_3	c_2	c_1	c_0
0	0	0	0	1
0	0	0	1	0
0	0	1	0	0
0	1	0	0	0
1	0	0	0	0

Since $\binom{r}{3} = \binom{5}{3} = 10$ is greater than $8(= k)$, any eight columns can be assigned to the information bits. Assuming

$$d_7 \quad d_6 \quad d_5 \quad d_4 \quad d_3 \quad d_2 \quad d_1 \quad d_0$$

are the information bits, a column assignment is selected for these bits, which, together with the previously assigned check bits, forms a parity check matrix for the (12, 8) Hsiao code, as shown below:

d_7	d_6	d_5	d_4	d_3	d_2	d_1	d_0	c_4	c_3	c_2	c_1	c_0	
0	1	1	0	1	1	0	0	0	0	0	0	1	e_0
1	0	1	1	0	1	1	0	0	0	0	1	0	e_1
1	1	1	0	0	0	1	1	0	0	1	0	0	e_2
1	0	0	1	1	0	0	1	0	1	0	0	0	e_3
0	1	0	1	1	1	1	1	1	0	0	0	0	e_4

From the parity check matrix it is easy to derive the equations for generating check bits:

$$c_0 = d_2 \oplus d_3 \oplus d_5 \oplus d_6$$
$$c_1 = d_1 \oplus d_2 \oplus d_4 \oplus d_5 \oplus d_7$$
$$c_2 = d_0 \oplus d_1 \oplus d_5 \oplus d_6 \oplus d_7$$
$$c_3 = d_0 \oplus d_3 \oplus d_4 \oplus d_7$$
$$c_4 = d_0 \oplus d_1 \oplus d_2 \oplus d_3 \oplus d_4 \oplus d_6$$

For example, the check bits for information bits

$$d_7 \, d_6 \, d_5 \, d_4 \, d_3 \, d_2 \, d_1 \, d_0 = 1\,0\,0\,0\,0\,0\,0\,1$$

are

$$c_4 \, c_3 \, c_2 \, c_1 \, c_0 = 1\,0\,0\,1\,0$$

Assuming bit d_1 is erroneous (i.e., 1), the check bits can be recomputed as

$$c_4' \, c_3' \, c_2' \, c_1' \, c_0' = 0\,0\,1\,0\,0$$

Therefore, the syndrome bits are

$$e_0 = c_0 \oplus c_0' = 0 \quad e_1 = c_1 \oplus c_1' = 1$$
$$e_2 = c_2 \oplus c_2' = 1 \quad e_3 = c_3 \oplus c_3' = 0$$
$$e_4 = c_4 \oplus c_4' = 1$$

It can be seen from the parity check matrix that the syndrome bits match with the seventh column from the right, identifying bit d_1 as the erroneous bit.

Next let us consider a double-bit error, for example, $d_3 d_2$ changes from 0 0 to 1 1. As a result the recomputed check bits become

$$c_4' \, c_3' \, c_2' \, c_1' \, c_0' = 1\,1\,0\,0\,0$$

The resulting syndrome bits are

$$e_0 \, e_1 \, e_2 \, e_3 \, e_4 = 0\,1\,0\,1\,0$$

Since this does not match with any column of the parity check matrix and the overall parity of the syndrome bits is 0, a double-bit error is indicated. Note that the syndrome bits match the EX-OR of columns d_3 and d_2 in the parity check matrix. The Hsiao code is an optimal code because it satisfies the condition of the minimum equal weight code [2.18]. The parity check matrix of such a code has the minimum number of 1s; also, the total number of 1s in each row is approximately equal.

2.5.3   Reed-Solomon Code

Reed-Solomon codes are nonbinary codes and are used in applications where multiple error corrections are needed. These codes are constructed and decoded using the *elements* of a finite field. A *finite field* consists of a set of elements and two operations (e.g., addition and multiplication). A field with q elements is called the *Galois field* GF(q). The number of elements in a Galois field must be of the form p^m, where p is a prime number and m is a positive integer. In cases where $p = 2$, all 2^m elements are derived using a polynomial $p(x)$ of degree m [2.19]. Polynomial $p(x)$ is selected such that the powers of a *primitive element* α up to $2^m - 2$ are distinct, and $\alpha^{2^m - 1} = 1$. Then $0, 1, \alpha, \alpha^2, \ldots, \alpha^{2^m} - 2, \alpha^{2^m - 1}$ is the set of 2^m field elements. Furthermore, each element in the field can be expressed as the sum of the elements $1, \alpha, \alpha^2, \ldots, \alpha^m - 1$ [2.20].

To illustrate let us consider a primitive polynomial with $m = 3$,

$$p(x) = x^3 + x + 1$$

Let α be a primitive element of the polynomial. Then

$$\alpha^3 + \alpha + 1 = 0$$

or

$$\alpha^3 = \alpha + 1$$

The elements of the field are

0

1

α

α^2

$\alpha^3 = \alpha + 1$

$\alpha^4 = \alpha(\alpha + 1) = \alpha^2 + \alpha$

$\alpha^5 = \alpha^3 + \alpha^2 = \alpha^2 + \alpha + 1$

$\alpha^6 = \alpha^3 + \alpha^2 + \alpha = \alpha + 1 + \alpha^2 + \alpha = \alpha^2 + 1$

$\alpha^7 = \alpha^3 + \alpha = \alpha + 1 + \alpha = 1$

A Reed-Solomon code (n, k) over $GF(2^m)$ can be used to encode a block of k data symbols each composed of m bits, to $n = 2^m - 1$ symbols. The code has a minimum distance $d = 2t + 1$, where $t (= \lfloor (n - k) \rfloor / 2)$ denotes the number of errors that can be corrected in a code word.

Let the data symbols to be encoded be represented by a polynomial

$$d(x) = d_0 + d_1 x + d_2 x^2 + \cdots + d_n x^n$$

The $(n - k)$ check symbols to be appended to the data symbols are the coefficients of the remainder polynomial $r(x)$ resulting from dividing $x^{(n-k)} \cdot d(x)$ by the generator polynomial $g(x)$:

$$g(x) = \prod_{i=1}^{d-1} (x + \alpha^i)$$

where α is a primitive element of $GF(2^m)$ [2.19].

	α^2	α	1
1	0	0	1
α	0	1	0
α^2	1	0	0
α^3	0	1	1
α^4	1	1	0
α^5	1	1	1
α^6	1	0	1

2.5

Table

Nonzero elements of $p(x) = x^3 + x + 1$.

To illustrate let us construct a (7, 3) Reed-Solomon code over $GF(2^3)$ with $d = 5$; thus, $t = 2$ errors can be corrected in this code. Table 2.5 shows the representations of seven nonzero elements of the primitive polynomial

$$p(x) = x^3 + x + 1$$

over $GF(2^3)$.

The generator polynomial for the (7, 3) Reed-Solomon code is

$$g(x) = \prod_{i=1}^{4}(x + \alpha^i)$$

$$= (x + \alpha^1)(x + \alpha^2)(x + \alpha^3)(x + \alpha^4)$$

$$= x^4 + \alpha^3 x^3 + x^2 + \alpha x + \alpha^3$$

Assuming the data to be encoded is

101 100 011

the corresponding data polynomial over $GF(2^3)$

$$d(x) = (\alpha^2 + 1)x^2 + \alpha^2 x + (\alpha + 1)$$

$$= \alpha^6 x^2 + \alpha^2 x + \alpha^3$$

The residual polynomial $r(x)$ resulting from the division of $x^{(7-3)} \cdot d(x)$ by $g(x)$ is

$$r(x) = \alpha x^2 + \alpha^5 x + 1$$

Therefore, the encoded data polynomial is

$$c(x) = \alpha^6 x^6 + \alpha^2 x^5 + \alpha^3 x^4 + \alpha x^2 + \alpha^5 x + 1$$

The code word can also be written as a vector

$$c(x) = (\alpha^6, \alpha^2, \alpha^3, 0, \alpha, \alpha^5, 1)$$

If converted to binary form the $(7, 3)$ Reed-Solomon code becomes a $(21, 9)$ code capable of double-3 bit error correction and double-3 bit error detection. The parity matrix of the code over $GF(2^3)$ is as follows [2.20]:

$$P = \begin{vmatrix} 1 & \alpha & \alpha^2 & \alpha^3 & \alpha^4 & \alpha^5 & \alpha^6 \\ 1 & (\alpha^2) & (\alpha^2)^2 & (\alpha^2)^3 & (\alpha^2)^4 & (\alpha^2)^5 & (\alpha^2)^6 \\ 1 & (\alpha^3) & (\alpha^3)^2 & (\alpha^3)^3 & (\alpha^3)^4 & (\alpha^3)^5 & (\alpha^3)^6 \\ 1 & (\alpha^4) & (\alpha^4)^2 & (\alpha^4)^3 & (\alpha^4)^4 & (\alpha^4)^5 & (\alpha^4)^6 \end{vmatrix}$$

$$= \begin{vmatrix} 1 & \alpha & \alpha^2 & \alpha^3 & \alpha^4 & \alpha^5 & \alpha^6 \\ 1 & \alpha^2 & \alpha^4 & \alpha^6 & \alpha & \alpha^3 & \alpha^5 \\ 1 & \alpha^3 & \alpha^6 & \alpha^2 & \alpha^5 & \alpha & \alpha^4 \\ 1 & \alpha^4 & \alpha & \alpha^5 & \alpha^2 & \alpha^6 & \alpha^3 \end{vmatrix}$$

To obtain the binary form, each element in the matrix is replaced by its corresponding binary matrix T_j defined as follows [2.20]:

$$\alpha^j \leftrightarrow T_j = \begin{vmatrix} | & | & & | \\ \alpha^j & \alpha^{j+1} & \cdots & \alpha^{j+m-1} \\ | & | & & | \end{vmatrix}$$

$$\text{for } j = 0, 1, \ldots, 2^m - 1$$

For example,

$$\alpha^3 \leftrightarrow T_3 = \begin{vmatrix} | & | & & | \\ \alpha^3 & \alpha^4 & \cdots & \alpha^5 \\ | & | & & | \end{vmatrix} = \begin{vmatrix} 1 & 0 & 1 \\ 1 & 1 & 1 \\ 0 & 1 & 1 \end{vmatrix}$$

Thus, the binary form of the parity matrix P is

$$\begin{vmatrix}
100 & 001 & 010 & 101 & 011 & 111 & 110 \\
010 & 101 & 011 & 111 & 110 & 100 & 001 \\
001 & 010 & 101 & 011 & 111 & 110 & 100 \\
& & & & & & \\
100 & 010 & 011 & 110 & 001 & 101 & 111 \\
010 & 011 & 110 & 001 & 101 & 111 & 100 \\
001 & 101 & 111 & 100 & 010 & 011 & 110 \\
& & & & & & \\
100 & 101 & 110 & 010 & 111 & 001 & 011 \\
010 & 111 & 001 & 011 & 100 & 101 & 110 \\
001 & 011 & 100 & 101 & 110 & 010 & 111 \\
& & & & & & \\
100 & 011 & 001 & 111 & 010 & 110 & 101 \\
010 & 110 & 101 & 100 & 011 & 001 & 111 \\
001 & 111 & 010 & 110 & 101 & 100 & 011
\end{vmatrix}$$

As discussed previously, the (7, 3) Reed-Solomon code has 3 bits in each symbol. If the number of bits in a symbol is 8 (i.e., a byte), then a code can have $255(= 2^8 - 1)$ symbols. Thus, the total number of bits in a code word will be $2040(= 255 \times 8)$. If t is selected to be 8, then

$$k = 255 - 2 \times 8 = 239$$

That is, there will be 239 data symbols or 1912 bits in a code word. Also, since the minimum Hamming distance between the code words is $17(= 2 \times 8 + 1)$, a byte error can be corrected in a code word. In other words, a (255, 239) Reed-Solomon code is a single-byte error correcting code.

Further discussion of the Reed-Solomon encoding and decoding can be found in [2.20].

2.6   REFERENCES

[2.1]   Sellers, F. W., M. Hsiao, and L. W. Bearnson. *Error Detecting Logic for Digital Computers.* McGraw-Hill, 1968.

[2.2]   Jha, N. K., and S. Wang. "Design and synthesis of self-checking VLSI circuits." *IEEE Trans. Computer-Aided Design* 12, no. 6 (1992): 878–887.

[2.3]   Sarrazin, D. B., and M. Malek. "Fault-tolerant semiconductor memories." *IEEE Computer* 17, no. 8 (1984): 49–56.

[2.4]   Mak, G. P., J. A. Abraham, and E. S. Davidson. "The design of PLAs with concurrent error detection." *Proc. 12th Int. Symp. on Fault-Tolerant Computing,* Milan, Italy, 1982, 202–210.

[2.5]   Wong, C. Y., W. K. Fuchs, J. A. Abraham, and E. S. Davidson. "The design of a microprogram control unit with concurrent error detection." *Proc. 12th Int. Symp. on Fault-Tolerant Computing*, Milan, Italy, 1982, 476–482.

[2.6]   Yen, M. M., W. K. Fuchs, and J. A. Abraham. "Designing for concurrent error detection in VLSI: Application to a microprogram control unit." *IEEE Journal of Solid-State Circuits,* SC-22, no.8 (1987): 595–605.

[2.7]   Smith, J.E. "The design of totally self-checking check circuits for a class of unordered code." *Jour. Design Automation and Fault-Tolerant Computing* 2 (October 1977): 321–343.

[2.8]   Bose, B., and T. R. N. Rao. "Theory of unidirectional error correcting/detecting codes." *IEEE Trans. Computers*, C-31, no. 6 (1982): 521–530.

[2.9]   Freiman, C. V. "Optimal error detection codes for completely asymmetric binary channels." *Information and Control* 5 (March 1962): 64–71.

[2.10] Berger, J. M. "A note on error detection codes for asymmetric binary channels." *Information and Control* 4 (March 1961): 68–72.

[2.11] Dong, H. "Modified Berger codes for detection of unidirectional errors." *IEEE Trans. Computers*, C-22 (June 1984): 572–575.

[2.12] Borden, J. M. "Optimal bound for asymmetric error correcting codes." *Information and Control* 52, no. 4 (1982): 66–72.

[2.13] Bose, B., and D. J. Lin. "Systematic unidirectional error detecting codes." *IEEE Trans. Computers*, C-34 (November 1985): 1026–1032.

[2.14] Berger, J. "A note on burst detecting sum code." *Information and Control* 4 (March 1961): 297–299.

[2.15] Bose, B. "Burst unidirectional error detecting codes." *IEEE Trans. Computers*, C-25 (April 1986): 250–252.

[2.16] Blaum, M. "On systematic burst unidirectional error detecting codes." *IEEE Trans. Computers* 27 (April 1988): 452–457.

[2.17] Hamming, R. W. "Error detecting and error correcting codes." *Bell Syst. Tech. Jour.* 26 (April 1950): 147–160.

[2.18] Hsiao, M. Y. "A class of optimal minimum odd-weight-column SEC-DED codes." *IBM Jour. Res. & Develop.* 14 (July 1970): 295–401.

[2.19] Lin, S. *An Introduction to Error-Correcting Codes.* Prentice Hall, 1970.

[2.20] Rao, T. R. N., and E. Fujiwara. *Error-Control Coding for Computer Systems.* Prentice Hall, 1989.

3 | Self-Checking Combinational Logic Design

$\mathbf{A}$s digital systems grow more and more complex, it is becoming highly desirable for maintenance and repair to have systems that have the capability of self-checking. *Self-checking* can be defined as the ability to verify automatically whether there is any fault in logic (chips, boards, or assembled systems), without the need for externally applied test stimuli. Thus, self-checking circuits allow *on-line error detection,* that is, faults can be detected during the normal operation of a circuit.

One way to achieve self-checking design is through the use of error detecting codes. The principle of such design can be better understood in terms of input/output mappings of the circuit. Let a circuit have m primary input lines and n primary output lines. Then the 2^m binary vectors of length m form the *input space X* of the circuit; the *output space Z* is similarly defined to be the set of 2^n binary vectors of length n. During normal, or fault-free, operation, the circuit receives only a subset of X called the *input code space* and produces a subset of Z called the *output code space.* Members of a code space are called *code words.* A non-code word at the output indicates the presence of a fault in the circuit. However, a fault may also result in an incorrect code word at the output, rather than a non-code word, in which case the fault is not detectable.

In general, a circuit may be designed to be self-checking only for an assumed set of faults. Such a set typically includes *single stuck-at faults* and *unidirectional multiple faults.* A single stuck-at fault assumes that a physical defect in a logic circuit results in one of the signal lines in the circuit being fixed to either a logic 0 (*stuck-at-0*) or logic 1 (*stuck-at-1*). If more than one

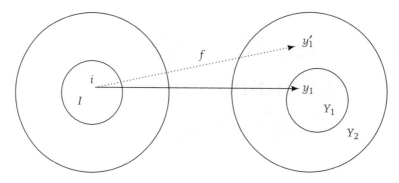

Self-testing property.

signal line in the circuit is stuck-at-1 or stuck-at-0 at the same time, the circuit is said to have a multiple stuck-at fault. A variation of the multiple fault is the unidirectional fault. A multiple fault is unidirectional if all of its constituent faults are either stuck-at-0 or stuck-at-1 but not both simultaneously.

Self-checking circuits must satisfy the following properties [3.1]:

+ Self-testing

+ Fault-secure

A circuit is self-testing if, for every fault from an assumed fault set, the circuit produces a non-code word at the output for at least one input code word. To illustrate let F be the assumed fault set. Also, let I and Y_1 be the set of input and output code words, respectively. As shown in Figure 3.1, if Z is the output space and Y_2 is the set of non-code words, then $Z = Y_1 \cup Y_2$. Let $i \in I$ and the correct output of the circuit be $y_1 \in Y_1$, and for the same input i the output of the circuit is $y_1' \in Y_2$ in the presence of a fault $f \in F$.

The self-testing property ensures that for every fault in the assumed set, there is at least one input code word for which the resulting output is a non-code word. In other words, the occurrence of any fault from the assumed set will be detected by at least one input code word.

A circuit is *fault-secure* for an assumed set of faults if, for any fault in the set, the circuit output is either a correct code word or a non-code word; that is, the circuit never produces an incorrect code word output for the input code words. In Figure 3.2 the mapping of an input code word to an output code word in the presence of a fault $f \in F$ is indicated by a solid line. Notice that the output in

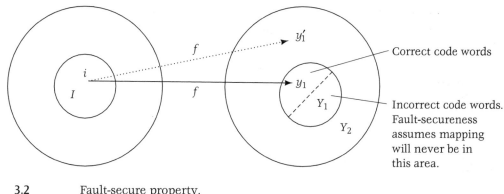

Correct code words

Incorrect code words. Fault-secureness assumes mapping will never be in this area.

3.2

Figure

Fault-secure property.

the presence of a fault cannot be a code word and at the same time be different from the correct code word output y_1. Thus, as long as the output is a code word, it must be assumed to be correct.

A circuit is *totally self-checking* if it is both self-testing and fault-secure. Totally self-checking circuits are very desirable for highly reliable digital system design, since during normal operation all faults from a given set would cause a detectable, erroneous output. Such circuits have many significant advantages [3.2], such as:

- ✦ Temporary faults as well as permanent faults are detected.

- ✦ Faults are immediately detected upon occurrence. This prevents corruption of data.

- ✦ Software diagnostic programs are eliminated or greatly simplified.

The model of a totally self-checking circuit is shown in Figure 3.3. It consists of a functional circuit and a checker (check circuit), both of which are totally self-checking. The function of the checker is to check the validity of the output code words. Thus, by observing the output of the checker, it is possible to detect any fault in the functional circuit or the checker itself. However, it is not possible to locate the fault (i.e., whether it is in the functional circuit or in the checker itself) from the information provided by the checker output.

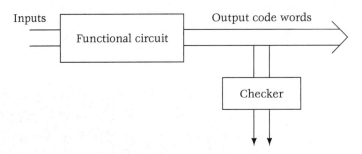

3.3

Figure

Model of a totally self-checking circuit.

3.1   STRONGLY FAULT-SECURE CIRCUITS

The effectiveness of totally self-checking circuits is based on the assumptions that each fault belongs to an assumed fault set, and faults occur sequentially with the time interval between two successive faults long enough to allow all code inputs to be applied to the circuit for testing. Thus, a totally self-checking circuit produces a non-code word as the erroneous output, if a fault is present.

If a circuit is fault-secure but not self-testing for fault set F, there could be a fault $f_1 \in F$ that does not produce non-code word outputs when code inputs are applied. In other words, fault f_1 is not detectable. When a second fault $f_2 \in F$ occurs, the circuit has a combined fault $f_1 \cup f_2$, which may not belong to F. Therefore, the circuit is not guaranteed to be fault-secure with respect to $f_1 \cup f_2$, and an incorrect code word may be produced.

A circuit is *strongly fault-secure* with respect to a set of faults F if, for a fault $f_1 \in F$, either

1. the circuit is self-testing and fault-secure or
2. the circuit is fault-secure for f_1, and if another fault $f_2 \in F$ occurs in the circuit, then either property 1 or 2 is true for the fault sequence $f_1 \cup f_2$ [3.3].

In general a circuit is strongly fault-secure with respect to a set of faults F if it is strongly fault-secure with respect to all fault sequences whose members belong to F [3.4]. It is easy to see that under this assumption any strongly fault-secure circuit achieves the totally self-checking goal. A circuit that is fault-secure but not strongly fault-secure can produce an incorrect code output before a non-code output.

Figure 3.4(a) shows a strongly fault-secure circuit [3.4]. It consists of three 4-bit registers R1, R2, and R3 that store 2-out-of-4 code words. Four 2-to-1

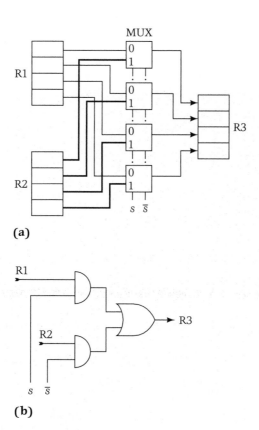

(a)

(b)

(a) A strongly fault-secure circuit; (b) a 2-to-1 multiplexer circuit.

multiplexers are used to transfer the contents of R1 or R2 to R3. The schematic of the 2-to-1 multiplexer is shown in Figure 3.4(b). If the control inputs $s = 0$ and $\bar{s} = 1$, then the content of R1 is transferred to R3; alternatively, if $s = 1$ and $\bar{s} = 0$, the content of R2 is stored in R3. In the presence of any sequence of faults from a set of unidirectional faults in the circuit, either a non-code word or the expected correct code word is stored in R3. Thus, this circuit is strongly fault-secure for the set of unidirectional faults.

3.2   STRONGLY CODE-DISJOINT CIRCUITS

A circuit is *code-disjoint* if input code words map into output code words, and non-code words into non-code word outputs [3.5]. The concept of code-disjoint

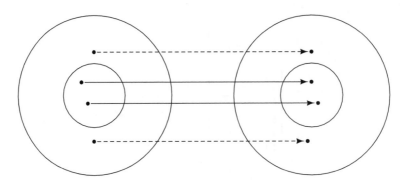

Code-disjoint property.

property is illustrated in Figure 3.5, where solid lines represent the mapping of input code words to output code words and dashed lines represent the mapping of non-code word inputs to non-code word outputs.

A circuit is *strongly code-disjoint* for an assumed fault set if, for every fault in the set, the circuit is either code-disjoint and self-testing or, if the circuit is not self-testing, it is still code-disjoint for a combination of an undetected fault and another fault from the fault set [3.6]. In other words, the circuit remains code-disjoint for any sequence of faults from the fault set. In certain cases, non-code word inputs map into code word outputs; this is known as *weakly code-disjoint* property [3.7]. Before discussing the properties of a weakly code-disjoint circuit, certain definitions need to be considered. A vector V *covers* another vector W if V has a 1 in every bit position in which W has a 1. If at least one code word covers a non-code word, the non-code word is called a *covered non-code word*. Alternatively, if a non-code word covers at least one code word, it is called a *covering non-code word*. Two vectors V and W are *unordered* if neither covers the other. A non-code word is said to be an *unordered non-code word* if it is unordered with all code words.

A circuit is considered to be weakly code-disjoint if and only if it satisfies the following:

1. It produces a non-code word output for an unordered non-code word input.

2. It produces a non-code output or the correct code output for a covering non-code word input i. In other words, the output is the same as that produced by the input code words covered by i.

3. It produces a non-code word output or a correct code word output for a covered non-code word input j; that is, the output code word is the same as that produced by input code words that cover j.

3.3   TERMINOLOGY

A *literal* is a Boolean variable or its complement. A *cube* is a set C of literals such that $x \in C$ implies $\bar{x} \notin C$. It is a conjunction of its literals (i.e., a product term). For example, if $\{\bar{a}, \bar{b}, \bar{c}\}$ is a cube, then it can be interpreted as the product term $\bar{a}\,\bar{b}\,\bar{c}$.

A *minterm* is a cube in which every variable of a Boolean function is present. A function in which each variable appears either in complemented or uncomplemented form, but not both, is *unate*.

The *distance* between input cubes C_i and C_j, denoted by dist (C_i, C_j), is the number of bit positions in which the cubes differ; the bit positions are either 1s or 0s but not don't cares. For example, if $C_i(a, b, c, d) = 1\,1\,0\,1$ and $C_j(a, b, c, d) = 0\,1 - 0$, dist $(C_i, C_j) = 2$ because the cubes differ in positions a and d.

Two output vectors O_i and O_j are *adjacent* if dist $(C_i, C_j) = 1$, where C_i and C_j are input cubes corresponding to O_i and O_j, respectively.

Two input cubes C_i and C_j are *bidirectionally adjacent* if their corresponding outputs O_i and O_j do not cover each other.

Two input cubes are *x-bidirectional* if they only differ in bit position x and their corresponding outputs are unordered. For example, input cubes C_1 and C_2 in Table 3.1 are *a-bidirectional*. A fault creates a *unidirectional error* if the correct and the faulty outputs are ordered. For example, if the output for cube C_1 changes to $0\,0\,0\,1$, it is covered by the correct output of C_1.

A primitive gate is *prime* if none of the inputs can be removed without changing the function of the resulting circuit. A gate is *irredundant* if its removal causes the resulting circuit to be functionally different. A gate-level circuit is said to be *prime* if all the gates are prime and *irredundant* if all the gates are irredundant. It has been shown [3.8] that a gate-level circuit is prime and irredundant if and only if it is 100% testable for all single stuck-at faults.

An undirected graph containing a set of vertices V and a set of edges E is denoted by G(V, E). G'(V', E') is a *subgraph* of G if V' is a subset of V, and an

	Inputs	Outputs
Cube	$a\ b\ c\ d$	$O_1\ O_2\ O_3\ O_4$
C_1	1 0 0 1	1 0 1 1
C_2	0 0 0 1	0 1 1 1

3.1

Table   Two bidirectional cubes.

edge joins two vertices in G′ if an edge joins the same two vertices in G. A *fully connected subgraph* of G is a graph $G_s(V_s, F_s)$ such that V_s is a subset of V, and any two vertices in G_s are connected by an edge.

A graph $G_1(V_1, E_1)$ covers graph $G_2(V_2, E_2)$ if the following are satisfied:

1. The number of vertices in E_1 is equal to or greater than the number of vertices in E_2.

2. If there exists an edge between two nodes in V_2, there should also be an edge between the same two nodes in V_1.

3.4   BIDIRECTIONAL ERROR-FREE COMBINATIONAL CIRCUIT DESIGN

Single stuck-at faults in combinational logic circuits can be detected if such a fault results in either a single-bit error or unidirectional multibit errors at the outputs, and the outputs are encoded using a single and unidirectional error detecting code. Therefore, if a combinational circuit can be designed such that all faults that result in bidirectional errors at the outputs are eliminated, the circuit will be totally self-checking [3.9].

Let us first illustrate with an example that it is possible to eliminate faults in a circuit that creates bidirectional errors. The truth table of a logic circuit that has four inputs and four outputs is shown in Table 3.2.[1] Notice that out of sixteen possible input combinations, only nine have been used in the truth table. These nine input combinations and their corresponding outputs have been arbitrarily encoded, with input combinations 0 0 0 1 and 0 1 1 0 having identical output 0 0 0 1.

Another encoding of the outputs with the same input encoding as in Table 3.2 is shown in Table 3.3. A third encoding that uses the same output encoding as in Table3.2 but uses 2-out-of-5 code for input encoding is shown in Table 3.4.

The circuits corresponding to the encodings of Tables 3.2–3.4 are shown in Figure 3.6(a), (b), and (c), respectively; these circuits are obtained by performing multilevel logic optimization on the encodings.

---

1. Tables 3.2 through 3.7 and Figures 3.6 through 3.9 are from [3.9], courtesy of Kluwer Academic Publishers, ©1994.

Inputs				Outputs			
a	b	c	d	Z_1	Z_2	Z_3	Z_4
0	0	0	1	0	0	0	1
0	0	1	0	0	0	1	0
0	0	1	1	0	1	0	0
0	1	0	0	1	0	0	0
0	1	0	1	1	1	0	0
0	1	1	0	0	0	0	1
0	1	1	1	1	0	0	1
1	0	0	0	0	1	1	0
1	0	0	1	0	1	0	1

3.2 Table Arbitrary input/output encoding.

Inputs				Outputs			
a	b	c	d	Z_1	Z_2	Z_3	Z_4
0	0	0	1	1	1	1	1
0	0	1	0	0	1	1	1
0	0	1	1	0	1	0	1
0	1	0	0	0	0	1	1
0	1	0	1	0	0	1	0
0	1	1	0	1	1	1	1
0	1	1	1	0	1	1	0
1	0	0	0	0	1	0	1
1	0	0	1	0	0	0	1

3.3 Table Output encoding.

Inputs					Outputs			
a	b	c	d	e	Z_1	Z_2	Z_3	Z_4
1	1	0	0	0	0	0	0	1
0	1	1	0	0	0	0	1	0
0	1	0	0	1	0	1	0	0
0	0	1	1	0	1	0	0	0
0	0	1	0	1	1	1	0	0
1	0	1	0	0	0	0	0	1
1	0	0	1	0	1	0	0	1
0	1	0	1	0	0	1	1	0
1	0	0	0	1	0	1	0	1

3.4 Table Input encoding.

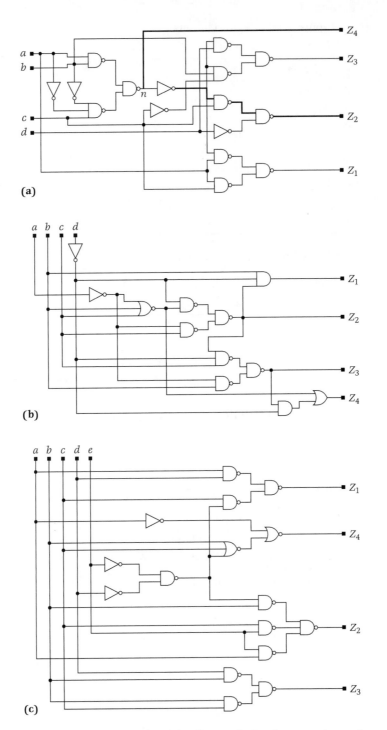

(a) Implementation using encoding 1; (b) implementation using encoding 2; (c) implementation using encoding 3.

Let us first consider the circuit shown in Figure 3.6(a). The output of the fault-free circuit corresponding to the input pattern $abcd = 0\,0\,0\,1$ is $Z_1 Z_2 Z_3 Z_4 = 0\,0\,0\,1$. If node n is stuck-at-0, the output changes to $Z_1 Z_2 Z_3 Z_4 = 0\,1\,0\,0$. The fault creates a bidirectional error at the outputs because the number of signal inversions between node n and output Z_4 is even, whereas the number of signal inversions between n and output Z_2 is odd.

The output encoding of Table 3.3 is done such that there are no m-bidirectional input cubes. Therefore, no fault in the circuit of Figure 3.6(b) can cause bidirectional error at the output because the number of signal inversions between any node in the circuit and any output is either odd or even, not a combination of the two.

A fault in the circuit of Figure 3.6(c), corresponding to the encoding shown in Table. 3.4, causes a single-bit error or unidirectional multibit errors at the outputs of the circuit.

3.5 DETECTION OF INPUT FAULT INDUCED BIDIRECTIONAL ERRORS

It should be clear from the above discussion that by properly encoding the inputs and/or the outputs of a circuit, bidirectional errors resulting from internal faults in a circuit can be eliminated. In this section we present a technique for identifying the presence of bidirectional errors in a circuit due to a fault in a primary input.

Let us assume that the two input cubes C_1 and C_2 of a circuit under test are x-bidirectional. Thus, C_1 and C_2, whose outputs O_1 and O_2 are partially bidirectional, differ in variable x. Therefore, a fault at x may activate C_2 when C_1 is applied, producing O_2 instead of O_1, thereby creating a bidirectional error.

If a circuit does not have any x-bidirectional input cubes, then either of the following two cases may be true:

Case 1 The distance between all bidirectionally adjacent input cubes is greater than or equal to two. ■

Case 2 Any adjacent outputs are ordered. ■

For Case 1 no fault at a primary input activates cube C_j instead of C_i where C_i and C_j are m-bidirectional; thus, any multiple error at the output will be of a unidirectional nature. For Case 2, a fault at a primary input may activate C_j

Cubes	$a\ b\ c$	O_1	O_2
C_1	1 0 0	1	0
C_2	0 1 0	1	1
C_3	0 0 1	1	0
C_4	1 1 1	1	0
C_5	1 1 0	1	1
C_6	0 1 1	0	1
C_7	1 0 1	0	1
C_8	0 0 0	0	1

3.5

Table

Truth table.

instead of C_i. Since the outputs corresponding to C_i and C_j are ordered, the fault results in single-bit or unidirectional multibit errors at the outputs.

The following algorithm identifies faults at primary inputs that may cause bidirectional errors at outputs [3.9]. The steps of the algorithm are listed below:

1. For every valid output O_i in a circuit, find all other outputs $O_j, O_k, \ldots,$ that are partially bidirectional with O_i. Use each of these outputs to form a group with O_i, such as $(O_i, O_j), (O_i, O_k), \ldots$ etc.

2. Find the distance between the input cube corresponding to output O_i (i.e., C_i) and the input cube corresponding to the other output in each output group. In other words, derive *dist* (C_i, C_j), *dist* $(C_i, C_k), \ldots$ etc.

3. If the distance between input cubes C_i and C_j is 1, and they differ in variable x, then add (C_i, C_j, x) to the set of possible bidirectional faults.

To illustrate the application of the algorithm, let us consider the truth table shown in Table 3.5. The application of the above algorithm to Table 3.5 results in the following set:

$\{(C_1, C_7, c), (C_1, C_8, a), (C_3, C_6, b), (C_3, C_7, a), (C_3, C_8, c), (C_3, C_7, a), (C_4, C_6, b)\}$

For example, the element in the set (C_1, C_7, c) indicates that if input cube C_1 is applied to the circuit designed according to the specification of Table 3.5, and input c is stuck-at-1, then the actual input will correspond to cube C_7. This will result in output 0 1 instead of the fault-free output 1 0.

3.6 TECHNIQUES FOR BIDIRECTIONAL ERROR ELIMINATION

Primary input faults that cause bidirectional errors at the outputs can be eliminated by encoding the inputs or the outputs such that no two input cubes become *m*-bidirectional [3.9].

3.6.1 Input Encoding

An input encoding approach assigns codes to two bidirectional input cubes, C_i and C_j, such that dist (C_i and C_j) ≥ 2. To illustrate let us consider a combinational circuit with I different input symbols and O encoded outputs. The required steps to satisfy the distance requirements are the following:

1. For each input cube in a circuit, derive all other input cubes that are bidirectionally adjacent to it.

2. Assign codes to each cube derived in Step 1, such that the distance between any two is greater than or equal to 2, by using the minimum number of encoding bits.

3. If an assignment is not possible, increase the number of encoding bits by one and repeat Step 2.

Let us illustrate the input encoding approach by applying it to the circuit specified in Table 3.6. Each input cube in the table is bidirectionally adjacent to four other cubes. This can be shown from the graph in Figure 3.7, which is constructed with input cubes as vertices and with an edge connecting two vertices if they are bidirectionally adjacent. Thus, the distance between any two vertices joined by an edge should be greater than or equal to 2. For example, cubes C_1, C_5, C_6, C_7, and C_8 is bidirectionally adjacent. If the minimum number of bits for encoding is used and C_1 is assigned $0\,0\,0$, then C_5, C_6, C_7, and C_8 have to be encoded as $1\,0\,1$, $1\,1\,0$, $0\,1\,1$, and $1\,1\,1$, respectively. Consequently, any code for C_2, C_3, or C_4 will have a distance of 1 from its bidirectionally

Input cubes	Output O_1	O_2
C_1	1	0
C_2	1	0
C_3	1	0
C_4	1	0
C_5	0	1
C_6	0	1
C_7	0	1
C_8	0	1

Table 3.6 Specification of a combinational circuit.

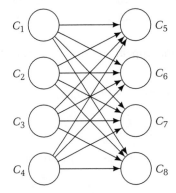

Figure 3.7 Graph corresponding to Table 3.6.

adjacent cubes, which is in conflict with the constraints in Step 2. Alternatively, an encoding with four bits as shown below satisfies the condition.

$$C_1 \rightarrow 0\,0\,0\,0 \qquad C_5 \rightarrow 1\,1\,0\,1$$
$$C_2 \rightarrow 0\,0\,1\,0 \qquad C_6 \rightarrow 1\,1\,1\,1$$
$$C_3 \rightarrow 0\,1\,0\,0 \qquad C_7 \rightarrow 1\,0\,0\,1$$
$$C_4 \rightarrow 0\,1\,1\,0 \qquad C_8 \rightarrow 1\,0\,1\,1$$

The proposed input encoding approach involves finding a Boolean cover that satisfies the distance requirement. However, finding an optimal Boolean cover is an NP-hard problem. If a heuristic approach is used instead, the cover is not guaranteed to be optimal.

An alternative approach for input encoding is to use m-out-of-n codes since the minimum distance between any two code words is at least 2. Also, the number of additional input lines needed for m-out-of-n encoding is not significant. In m-out-of-n encoding a stuck-at-0 (stuck-at-1) fault at an input line can only cause a transition from 1 to 0 (0 to 1) at the outputs if only the true values of the input variables are considered. Therefore, if input C_i covers input C_j, then O_i (the output for C_i) is either equal to O_j (the output for C_j) or for O_i covers O_j. If a non-ordered input C_n is covered by a set of code words, each of which has output O_n, then the output of C_n will also be O_n after minimization. For example, let us assume that a non-code word $1\,0\,0\,0$ is covered by code words $1\,1\,0\,0$, $1\,0\,1\,0$, and $1\,0\,0\,1$, and the output of the code words is 1. Then the output of non-covered $1\,0\,0\,0$ will also be 1. On the other hand, if the output of one of the code words that covers $1\,0\,0\,0$ is 0, the output of C_n will also be 0 because $1\,0\,0\,0$ cannot be used for the minimization of the function.

If a circuit is designed such that the output functions are unate, then variables in the output expressions appear either in normal or complement form but not both. Thus, if a variable appears in a noncomplement form in an expression, the variable either does not get inverted or goes through an even number of inversions from the input to the output. Similarly, if a variable appears in a complement form, it goes through an odd number of inversions. Therefore, if a fault occurs at an input line of a circuit that is implemented from a unate expression, then the fault will create a single-bit error if only one output is affected or a unidirectional error if more than one output is affected. Note that if a fault occurs at an internal node x rather than at a primary input of the circuit and affects k outputs, then there will be at least k different paths from node x to k affected outputs.

The number of inversions from primary input i to any output O_k is the same. This number is the sum (mod 2) of the number of inversions from primary input i to node x, and the number of inversions from node x to output O_k. This implies that the number of inversions from node x to outputs O_k is the same. Therefore, a fault at node x can create only a unidirectional error since the number of inversions from the fault site to the output is either odd or even.

Based on the above discussion, it can be concluded that if an input variable is present in output expressions of a circuit in either normal or complement form but not both, then any fault located in a path leading from the primary input corresponding to the input variable to the primary outputs will cause either a single-bit error or unidirectional multibit errors at the outputs. Thus, no fault in such a circuit will produce bidirectional errors.

3.6.2   Output Encoding

An alternative approach to eliminate bidirectional errors is to encode outputs corresponding to two input cubes C_i and C_j, with dist $(C_i, C_j) = 1$, such that they are ordered. To illustrate, let us consider Table 3.7, in which Z_1, Z_2, and Z_3 are the symbolic outputs. If the outputs are encoded as shown in the table, then C_1 and C_3 are x-bidirectional. However, by changing the encoding of Z_3 from 0 1 to 1 1, it can be guaranteed that a fault on primary input line x will not produce bidirectional errors.

An algorithm for encoding the N symbolic outputs of a circuit so that there are no m-bidirectional input cubes is given below.

1. Initialize m to $\lceil \log N \rceil$.

2. Construct an undirected graph G with N vertices.

3. Connect vertex O_i to vertex O_j if outputs O_i and O_j are adjacent.

4. Construct an undirected graph G_m with 2_m vertices.

5. If G_m covers G, then an output encoding assignment is obtained from G_m; exit. Otherwise, increment m by 1 and go to Step 4.

To illustrate the application of the algorithm, let us consider the specification of a circuit shown in Table 3.8. Since $N = 4$, graph G has four vertices as shown in Figure 3.8. Let us construct graph G_3 with 2^3 vertices as shown in Figure 3.9. The number of edges for a vertex depends on the number of 1s in

| | Inputs | | | | |
Cube	x	y	z	Symbolic outputs	Encoded outputs
C_1	0	1	0	Z_1	1 0
C_2	0	0	1	Z_2	0 0
C_3	1	1	0	Z_3	0 1

3.7   Symbolic and encoded outputs for a set of cubes.

Table

| | Inputs | | | Outputs |
Input cubes	x	y	z	f
C_1	0	0	1	O_1
C_2	1	–	1	O_2
C_3	0	1	1	O_3
C_4	–	–	0	O_4

3.8

Table

Specification of a combinational circuit.

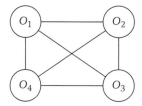

3.8

Figure

Graph with four vertices.

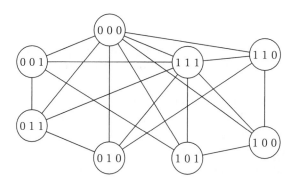

3.9

Figure

Graph G_3.

an m-bit vertex. If k is the number of 1s in an m-bit vertex, then the number of edges for that vertex is

$$(2^k - 1) + (2^{(m-k)} - 1)$$

Any two vertices in the graph are connected if one covers the other. Since vertex 1 1 1 covers, and vertex 0 0 0 is covered by, all other vertices, these two vertices are connected to all others. Graph G is fully connected and is covered by G_3. A graph Y is covered by another graph X if the number of vertices in X is equal to or greater than the number of vertices in Y, and if there exists an edge between two nodes in Y there should also be an edge between the same two nodes in X.

A fully connected subgraph of G_3 containing four vertices is needed for encoding the outputs of Figure 3.8. Any of the subgraphs shown in Figure 3.10 can be used to encode the outputs. Therefore, the outputs can be encoded as below:

O_1	0 0 0	0 0 0	0 0 0	0 0 0
O_2	1 1 1	1 1 1	1 1 1	1 1 1
O_3	0 1 0	1 0 0	1 0 1	1 1 0
O_4	0 1 1	1 0 1	0 0 1	1 0 0

The drawback of the output encoding algorithm is that Step 5 is a graph embedding problem; the solving of Step 5 by using the smallest m requires an exhaustive search. A heuristic solution with polynomial complexity can be found in [3.9].

If the outputs of a prime irredundant combinational circuit are encoded using the above algorithm, all m-directional input cubes are eliminated. Therefore, all single faults at primary input lines can cause single or unidirectional errors at the output. Also, since the circuit is prime and irredundant and hence fully testable, there exists at least one input pattern that creates a path from an input through the node to an output. If for an input pattern there is only one such path from an input to an output through node x, a fault at node x will produce a single-bit error. On the other hand, if there are k sensitizable paths from a primary input through node x to k different outputs, then a change in input will change the k outputs. Since any input change cannot cause a bidirectional change at the outputs because of the way the outputs are encoded, a fault at node x cannot cause bidirectional errors. Consequently, a single stuck-at fault causes either a single-bit error or unidirectional errors at outputs.

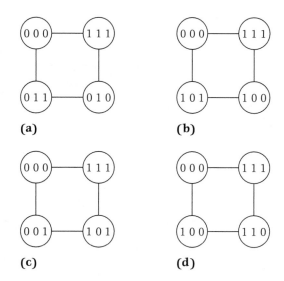

(a) **(b)**

(c) **(d)**

3.10

Figure

Subgraphs of G_3 with four vertices.

3.7 SELF-DUAL PARITY CHECKING

Sellers et al. [3.10] have proposed a method of on-line error detection in combinational circuits based on a modification of the conventional parity checking. In conventional parity checking, as shown in Figure 3.11, the parity bit p corresponding to the output bits of the combinational circuit is compared with the parity bit p' generated independently by the parity prediction circuit. Assuming the combinational circuit implements the following functions

$$y_1 = f_1(x_1, x_2, \ldots, x_n)$$
$$y_2 = f_2(x_1, x_2, \ldots, x_n)$$

$$\vdots$$

$$y_n = f_n(x_1, x_2, \ldots, x_n)$$

the parity prediction function y_p is defined by

$$y_p = y_1 \oplus y_2 \oplus \ldots \oplus y_n$$

In general the area overhead for separate implementation of the parity prediction circuit—that is, no shared logic between it and the combinational circuit—results in an average of 33% area overhead of the combinational circuit [3.11].

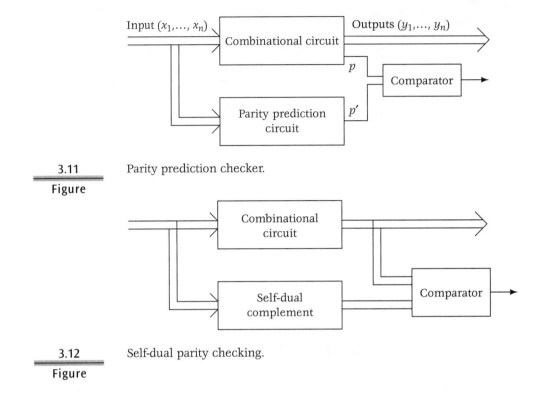

3.11

Figure Parity prediction checker.

3.12

Figure Self-dual parity checking.

Saposhnikov et al. [3.12] have also proposed a method for on-line error detection that uses a modified form of conventional parity checking to reduce area overhead. In this method the parity prediction circuit of Figure 3.11 is replaced by an alternative circuit that generates a self-dual complement of the Boolean function implemented by the combinational circuit. Figure 3.12 shows the configuration used by the proposed scheme.

A Boolean function $f(x)$ is *self-dual* if it is equal to its dual function $f_d(x)$. For example, the dual of the function

$$f(a, b, c) = \bar{a}\,\bar{c} + b\bar{c} + \bar{a}b$$

is

$$f_d(a, b, c) = (\bar{a} + \bar{c})(b + \bar{c})(\bar{a} + b)$$

It is easy to show that after expansion f_d becomes equal to f; thus, f is self-dual.

It is possible to determine whether an *n*-variable function f is self-dual without actually deriving f_d and verifying it is equal to f, by checking the following two properties in the sum-of-products representation of f [3.13]:

✦ it has exactly 2^{n-1} minterms
✦ for each $m_i \in f$, $m_{2^n - 1 - i} \in \overline{f}$

The self-dual complement of function $f(x)$, if one exists, must satisfy the following:

$$h(x) = f(x) \oplus \delta(x)$$

where $\delta(x)$ is the self-dual complement of $f(x)$, and $h(x)$ is an arbitrary self-dual function that is the linear sum of $\delta(x)$ and the output of the circuit corresponding to function $f(x)$.

To illustrate, let us derive the self-dual complement of the function

$$f(a, b, c, d) = a\overline{c} + \overline{c}d + ad$$

Since $h(a, b, c, d)$ can be an arbitrary self-dual function, a possible choice is

$$h(a, b, c, d) = \overline{a}b\overline{c} + a\overline{b}\overline{c}\overline{d} + abd + \overline{a}bcd + bc\overline{d}$$

This expression can be rewritten as

$$h(a, b, c, d) = f(a, b, c, d) \oplus (\overline{b}d + b\overline{d})$$

Thus, the self-dual complement of $f(a, b, c, d)$ is $\overline{b}d + b\overline{d}$.

The implementation of $f(a, b, c, d)$ and its self-dual complement is shown in Figure 3.13.

For a circuit with multiple outputs such as $y_1, y_2, \ldots, y_n$, the parity of the output bits is compared with the self-dual complements of the parity function. In other words, if f_p is the parity function, that is,

$$f_p = y_1 + y_2 + \ldots + y_n \text{ where } y_i \ (i = 1, \ldots, n) = f_i(x_1, x_2, \ldots, x_m)$$

then δ_p, the self-dual complement of f_p, is chosen such that the function

$$h(x_1, \ldots, x_m) = f_p(x_1, \ldots, x_m) \oplus \delta_p(x_1, \ldots, x_m)$$

is self-dual. Figure 3.14 shows the implementation of $h(x)$.

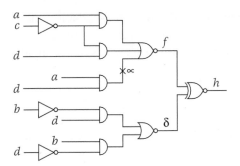

3.13

Figure

Implementation of a function and its self-dual complement.

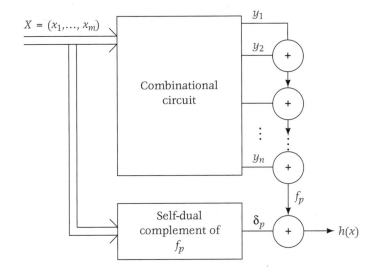

3.14

Figure

Self-dual parity checking of multi-output circuits.

During normal operation complementary input patterns are applied to the composite circuit implementing $h(x)$ at time units t and $t + 1$. If there is no fault in the self-complement dual circuit or in the functional circuit, the resulting output responses are also complementary. Thus, the self-dual parity checking, in conjunction with the time redundancy scheme (see Section 6.3), allows on-line error detection in the circuit corresponding to function $h(x)$.

To illustrate let us assume that node α is stuck-at-0 in Figure 3.13. The responses of the circuit to input pattern 1 1 1 1 and its complement 0 0 0 0 are shown below:

a	b	c	d	h
1	1	1	1	1
0	0	0	0	0

no fault

a	b	c	d	h
1	1	1	1	0
0	0	0	0	0

α stuck-at-0

The drawback of the self-dual parity check approach is that the derivation of an optimal self-dual complement of a given Boolean function may require an excessive amount of computation time. Another disadvantage is that it requires 100% time redundancy, in addition to hardware overhead that results from the self-dual complement circuit.

3.8   SELF-CHECKING DESIGN USING LOW-COST RESIDUE CODE

The *mod-3 residue code,* also known as *low-cost residue code,* can be used in devices such as multipliers, ALUs, and the like for on-line error detection of permanent or transient errors [3.14]. The main features of mod-3 residue codes are as follows:

1. There are only two check bits in a code word.
2. The residue of bit i in a binary pattern—that is, 2^i (mod 3)—is 1 if bit i occupies an even bit position and 2 if bit i occupies an odd bit position.

For example, the residue of the binary pattern 1 0 1 0 1 1 is $(2 + 0 + 2 + 0 + 2 + 1)$ mod $3 = 0\,1$, assuming the rightmost bit is the least significant bit. Thus, the resulting code word is 1 0 1 0 1 1 0 1.

To understand how low-cost residue code can be used in self-checking, let us assume that the outputs of a circuit are encoded using low-cost residue code, and z and z' are the decimal equivalents of the fault-free and the faulty outputs. Thus,

$$z - z' = \pm w$$

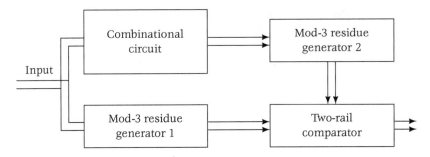

3.15

Figure

Mod-3 residue code–based self-checking circuit.

If $w \bmod 3 \neq 0$, a single-bit or a multibit error is present in the outputs. However, $w \bmod 3 = 0$ does not necessarily imply that the outputs are fault-free. This is because w could be a multiple of 3 even though z and z' are not equal, thus masking the presence of errors in the output pattern. It should be clear from the properties of low-cost residue code that only a multibit error at the output pattern can be masked.

To illustrate let us assume that the output pattern of a combinational circuit is

$$y_3\ y_2\ y_1\ y_0 = 1\ 0\ 1\ 1$$

Thus, the residue of the output pattern is 2 ($= 1\ 0$ in binary). Next assume that a fault in the circuit creates a single-bit error in the output:

$$y_3\ y_2\ y_1\ y_0 = 1\ 0\ 1\ 0$$

The residue of the faulty output pattern is 1 ($0\ 1$ in binary). Since the residues of the actual and the faulty output patterns are different (i.e., $w = \pm 1$), the fault is detectable.

On the other hand, if a fault creates a multibit error at the outputs, the residues of the faulty and the fault-free outputs may be identical. For example, if a fault changes the output pattern to

$$y_3\ y_2\ y_1\ y_0 = 1\ 0\ 0\ 0$$

the residue will be 2 (i.e., $w = 0$) and hence the fault cannot be detected. Thus, the encoding of outputs using mod-3 residue can guarantee that a circuit will be self-checking for single-bit errors; however, certain combinations of multibit errors at the outputs may not be detected.

Figure 3.15 shows the basic scheme for implementing self-checking combinational circuits using mod-3 residue code. The residue generator 1 is used to generate the predicted residue of an output pattern from the input pattern producing the output. The residue generator 2 generates the complement of the residue directly from the output pattern. A *two-rail checker* (see Chapter 4) compares the two residues and produces a non-code word (0 0 or 1 1) if the two residues are not complements of each other.

3.9   TOTALLY SELF-CHECKING PLA DESIGN

The PLA (programmable logic array) is used for the implementation of combinational logic functions in VLSI chips. Although PLAs in general need larger chip areas than the random logic, they have the advantage of a memorylike regular structure. The PLA logic structure is shown in Figure 3.16(a). The input lines to the AND array are called *bit lines,* and the outputs of the AND array are called *word lines* or *product lines.* The input signals and their complements enter the AND array and are selectively connected to product lines in such a way that certain combinations of input variables produce a logic 1 signal on one or more product lines. The product lines are input to the OR array; hence the outputs of the OR array are the sum-of-products form of Boolean functions of the PLA inputs. A cross-point b_k, p_k is marked ($\bullet$), if the product term p_k depends on b_k, and a cross-point z_j, p_k is marked if p_k is a term of z_j. Figure 3.16(b) shows an example of a PLA realizing the functions

$$z_1 = x_1 x_2 + \overline{x}_1 \overline{x}_2$$
$$z_2 = \overline{x}_1 x_2$$

Three kinds of faults can normally occur in PLAs: stuck-at faults, bridging faults, and cross-point faults [3.15, 3.16, 3.17]. A cross-point or contact fault may arise due to a missing contact on the cross-point of the AND and the OR arrays where there should be a contact, or due to an additional contact at the cross-point where there should not be a contact.

Wang and Avizienis [3.18] were the first to propose an approach for designing totally self-checking PLAs. Their approach is based on the assumption that the PLA is "non-concurrent"; that is, any normal input pattern selects exactly one product term in the PLA under fault-free operation, and the output patterns of the PLA belong to a code for which a totally self-checking checker exists. However, they did not mention the effect of different types of faults. Also, if the PLAs do not produce all the possible output combinations, this approach does not guarantee that all faults in checker circuits will be detected; hence the checkers are not totally self-checking.

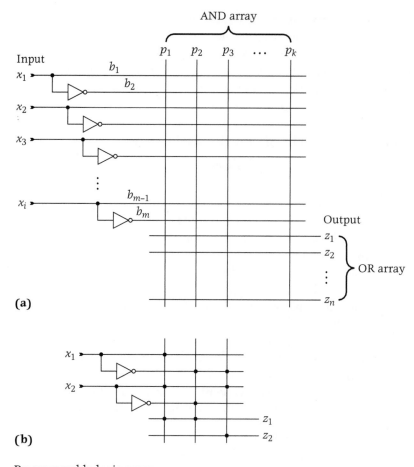

(a)

(b)

3.16

Figure

Programmable logic array.

Mak et al. [3.19] have shown that all single faults in a PLA can cause only unidirectional errors in the output of the PLA. Consequently, codes that can detect unidirectional errors such as *m*-out-of-*n* codes, Berger codes, or two-rail codes may be used for designing totally self-checking PLAs. Mak et al. [3.19] also introduce a new type of code called *modified Berger code,* specifically for detecting unidirectional output errors in PLAs. The modified Berger code requires fewer check bits than the original Berger code. The check bits for the modified Berger code are the binary encoding of the difference between the number of 0s and the minimum number of 0s in the output patterns.

A general procedure for designing strongly fault-secure (see Section 3.1) PLAs has also been introduced [3.19]. The first step of the procedure is to select an appropriate code to encode the outputs and to determine the assignment for check bit values. The check bit functions for the selected code are then derived. The next step is to join together the check bit function to the original output function to form a new multi-output function. The new function is then minimized by using a multi-output function minimization method to generate a minimal set of product terms. Finally, the AND and the OR arrays are implemented using the previously selected coding scheme.

As an example, let us consider how the procedure for strongly fault-secure PLA design can be applied to the following four-input, four-output functions:

$$f_1(x_1, x_2, x_3, x_4) = \sum(0, 2, 3, 7, 8, 10, 12, 13, 15)$$

$$f_1(x_1, x_2, x_3, x_4) = \sum(0, 2, 3, 4, 9, 12, 13, 15)$$

$$f_1(x_1, x_2, x_3, x_4) = \sum(0, 1, 2, 4, 8, 9, 10, 14)$$

$$f_1(x_1, x_2, x_3, x_4) = \sum(0, 1, 2, 4, 5, 6, 8, 11, 14)$$

Since the number of outputs is four, the extra output bits required for Berger coding are three. The modified Berger code, on the other hand, requires only two extra output bits, as can be deduced from Figure 3.17(a), (b), and (c). The assignment of check bit values for the modified Berger encoding is shown in Figure 3.18. Since the entry in row 1 and column 1 of Figure 3.17(c) is 0 (i.e., $f_1 f_2 f_3 f_4 = 1\,1\,1\,1$), the number of check bits to be assigned is $C_1 C_2 = 0\,0$. Hence the entries in the first row and the first column of both C_1 and C_2 are 0s.

The encoded truth table for the four-input, four-output function is shown in Table 3.9.

The check bit functions C_1 and C_2 are added to the original functions f_1, f_2, f_3, and f_4; hence the PLA would have six output lines. Using a multi-output minimization algorithm [3.20], the functions $f_1, \ldots, f_4, C_1$, and C_2 are minimized to generate a set of 14 product terms:

$$P_1 = \overline{x}_1\overline{x}_2 x_3 x_4 \quad P_5 = \overline{x}_1 x_2 x_3 x_4 \quad P_9 = x_1\overline{x}_2\overline{x}_3 x_4 \quad P_{13} = \overline{x}_1 x_2 x_3 \overline{x}_4$$

$$P_2 = x_1 x_2 \overline{x}_3 \quad P_6 = \overline{x}_1 \overline{x}_2 \overline{x}_4 \quad P_{10} = \overline{x}_1 \overline{x}_2 \overline{x}_3 x_4 \quad P_{14} = x_1 \overline{x}_2 x_3 x_4$$

$$P_3 = x_1 x_2 x_4 \quad P_7 = x_1 \overline{x}_2 \overline{x}_3 \overline{x}_4 \quad P_{11} = x_1 x_2 x_3 \overline{x}_4$$

$$P_4 = x_1 \overline{x}_2 x_3 \overline{x}_4 \quad P_8 = \overline{x}_1 x_2 \overline{x}_3 \overline{x}_4 \quad P_{12} = \overline{x}_1 x_2 \overline{x}_3 x_4$$

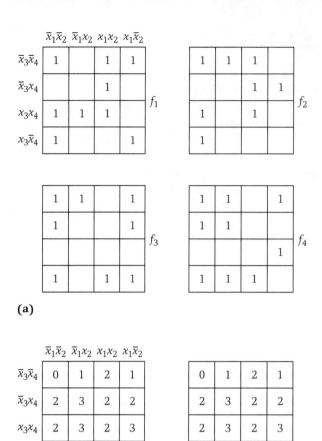

(a)

(b)

	$\bar{x}_1\bar{x}_2$	$\bar{x}_1 x_2$	$x_1 x_2$	$x_1\bar{x}_2$
$\bar{x}_3\bar{x}_4$	0	1	2	1
$\bar{x}_3 x_4$	2	3	2	2
$x_3 x_4$	2	3	2	3
$x_3\bar{x}_4$	0	3	2	2

(c)

0	1	2	1
2	3	2	2
2	3	2	3
0	3	2	2

3.17
Figure

(a) Karnaugh maps for a four-input, four-output function; (b) number of 0s in the output pattern corresponding to each input combination; (c) difference of the number of 0s and the minimum number of 0s.

	$\bar{x}_1\bar{x}_2$	$\bar{x}_1 x_2$	$x_1 x_2$	$x_1\bar{x}_2$
$\bar{x}_3\bar{x}_4$			1	
$\bar{x}_3 x_4$	1	1	1	1
$x_3 x_4$	1	1	1	1
$x_3\bar{x}_4$		1	1	1

C_1

	$\bar{x}_1\bar{x}_2$	$\bar{x}_1 x_2$	$x_1 x_2$	$x_1\bar{x}_2$
$\bar{x}_3\bar{x}_4$	1			1
$\bar{x}_3 x_4$	1			
$x_3 x_4$	1			1
$x_3\bar{x}_4$	1			

C_2

3.18
Figure

Check bit assignment.

x_1	x_2	x_3	x_4	f_1	f_2	f_3	f_4	C_1	C_2
0	0	0	0	1	1	1	1	0	0
0	0	0	1	0	0	1	1	1	0
0	0	1	0	1	1	1	1	0	0
0	0	1	1	1	1	0	0	1	0
0	1	0	0	0	1	1	1	0	1
0	1	0	1	0	0	0	1	1	1
0	1	1	0	0	0	0	1	1	1
0	1	1	1	1	0	0	0	1	1
1	0	0	0	1	0	1	1	0	1
1	0	0	1	0	1	1	0	1	0
1	0	1	0	1	0	1	0	1	0
1	0	1	1	0	0	0	1	1	1
1	1	0	0	1	1	0	0	1	0
1	1	0	1	1	1	0	0	1	0
1	1	1	0	0	0	1	1	1	0
1	1	1	1	1	1	0	0	1	0

3.9

Table

Coded truth table.

The output functions of the PLA can now be expressed as a sum of the product terms:

$$f_1 = P_1 + P_2 + P_3 + P_4 + P_5 + P_6 + P_7$$
$$f_2 = P_1 + P_2 + P_3 + P_6 + P_8 + P_9$$
$$f_3 = P_4 + P_6 + P_7 + P_8 + P_9 + P_{10} + P_{11}$$
$$f_4 = P_6 + P_7 + P_8 + P_{10} + P_{11} + P_{12} + P_{13} + P_{14}$$
$$C_1 = P_1 + P_2 + P_3 + P_4 + P_5 + P_9 + P_{10} + P_{11} + P_{12} + P_{13} + P_{14}$$
$$C_2 = P_5 + P_7 + P_8 + P_{12} + P_{13} + P_{14}$$

The number of product terms required for implementing strongly fault-secure PLAs is somewhat larger than that for conventional PLAs; in the example considered here, only one extra product term was required. The PLA outputs have been encoded into the modified Berger code. Any single stuck-at fault, bridging fault, or contact fault in the PLA will cause a unidirectional error at the output; an output word with a unidirectional error is a non-code word. If

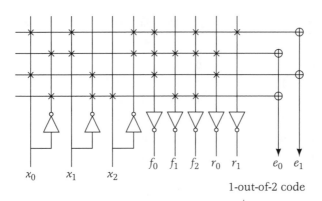

f_0 f_1 f_2 r_0 r_1 e_0 e_1

x_0 x_1 x_2

1-out-of-2 code

3.19   Self-checking PLA.

Figure

a fault does not cause a unidirectional error at the output, the output code word is correct and the fault is undetectable. The checker needed to detect non-code words at the PLA output must be designed so that the output produced by the PLA is sufficient to detect all faults in the checker.

An alternative approach for totally self-checking PLA design uses a low-cost residue code with mod 3 and an EX-OR tree [3.21]. If the output of a PLA is encoded using a low-cost residue code with check base 3, then any single-bit error at the output can be detected. The problem with encoding output patterns alone is that if a fault produces multibit errors at the output, the resulting mod-3 residue may be the same as in the error-free case, thus preventing the detection of output errors. However, by incorporating two EX-OR trees such that alternate product terms in the PLA are inputs to each EX-OR tree, it is possible not only to detect all single-bit errors at the output but the majority of multibit errors as well.

This self-checking design technique is illustrated using the PLA shown in Figure 3.19. First, the PLA has to be designed so that only one product term is enabled for an input pattern. Thus, in the presence of a fault, a product term may not be activated, or an additional product term together with the desired product term may be activated. As will be clear from Figure 3.19, if only one product term is activated, the outputs of the EX-OR trees, $e_0 e_1$, will be either 0 1 or 1 0. If none of the product terms is activated, $e_0 e_1$ will be 0 0, whereas if an undesired product term is activated in addition to the desired one, $e_0 e_1$ will be 1 1. Thus, in either case $e_0 e_1$ does not form 1-out-of-2 code, hence the fault will be detected on-line.

The output of the PLA is encoded using mod-3 residue code, and therefore two extra output lines, r_0 and r_1, are included. As can be seen in Figure 3.19, the residue of each output pattern is represented by these extra lines. For example, the sum lines f_0, f_1, and f_2 in the first row of Figure 3.19 constitute the decimal number 5. The mod-3 residue of 5 is 2, thus $r_0r_1 = 0\,1$. Therefore, the spurious presence or absence of a product term will be indicated by a residue value that is different from the expected one.

In summary, the above technique allows the detection of single faults in product terms by the EX-OR trees and in sum lines by the mod-3 residue codes. Therefore, all single faults in a PLA can be detected. A significant portion of multibit errors in output patterns can also be detected by this technique.

3.10   FAIL-SAFE COMBINATIONAL CIRCUIT DESIGN

Fail-safe realization of digital circuits is extremely important in applications where a fault can have catastrophic consequences. In a fail-safe system the output assumes one of the predetermined safe values when a fault occurs. The concept of fail-safe operation was first proposed by Mine and Koga [3.22]. A circuit is considered to be fail-safe if in the presence of a fault the circuit produces either the correct output or a predetermined safe output such as all 0s or all 1s. Thus, a fault in a fail-safe circuit forces the circuit to produce safe output so that there are no damaging consequences in the presence of a fault in the circuit. For example, a fail-safe traffic light system in the presence of a fault should remain at red. It should be pointed out that the erroneous outputs of a fail-safe circuit must belong to the expected output set of the circuit. Nicolaidis [3.23] has proposed that a circuit is fail-safe with respect to a fault if, in the presence of a fault from the set, the output of the circuit is either correct or safe. Thus, if X is the input space for the circuit, F is the fault set, C is the correct output set of the circuit, and S is the safe output set, then the circuit is fail-safe if, in the presence of fault $f \in F$, $x \in X$ maps into $c \in C$ or $s \in S$.

Analogous to the totally self-checking goal for self-checking circuits, a totally fail-safe goal may be achieved by ensuring that all erroneous outputs produced by a circuit before the fault detection process is started belong to the safe output set. As in the case of a totally self-checking circuit, it is assumed that faults occur one at a time, and all input combinations necessary to detect all detectable faults in the circuit can be applied between the occurrence of any two faults from the assumed fault set. Clearly, a fail-safe system should be such that during normal operation it does not produce nonsafe outputs; that is, all erroneous outputs should belong to the safe set. A fail-safe system that satisfies this goal is categorized as being *ultimate fail-safe* [3.23]. An ultimate fail-safe system has to be designed such that a fault in a system that results in a detectable error must force the system into an irreversible safe state; the system should remain in this state irrespective of any other fault that may occur in the system. For example, a totally self-checking checker normally produces 1-out-of-2 code (see Chapter 4); in order to be ultimate fail-safe, the checker must move into an erroneous but safe output state if there is a fault in the functional or in the error indicator circuit.

To illustrate let us design a totally self-checking checker for 2-out-of-4 code that is also ultimate fail-safe. The checker circuit shown in Figure 3.20 is derived from Table 3.10.

Let us assume that $0\,0$ and $1\,1$ are the safe and the nonsafe erroneous output, respectively. To ensure that the checker is ultimate fail-safe, a translator circuit is designed such that if the checker produces the erroneous output $1\,1$, the output of the translator is locked into 0; no subsequent fault in the circuit can change this state. The translator is designed using pseudo-nMOS circuit style and is shown in Figure 3.21. The output s of the translator circuit remains at 1 if the checker produces either a correct or the safe erroneous output ($0\,0$). If the checker produces $1\,1$ (i.e., the erroneous and nonsafe output), node A in the translator circuit is tied to 0, resulting in a 1 at node B. Since output s is normally held at 1, logic 1 at node B is transferred to node C (slightly degraded value). This turns on the n-transistor connected to output node s, resulting in a 0 at the output. As a result the p-transistor connected to node C is turned on, which in turn keeps the n-transistor connected to s permanently on. Thus, the output of the translator is locked in a state of 0.

To illustrate the design of an ultimate fail-safe circuit, let us consider the truth table of a function with three inputs (a, b, c) and two outputs (x, y) as shown below:

a b c	x y	
0 0 0	0 0	—— safe output
0 0 1	0 1	
0 1 0	0 1	
0 1 1	1 0	valid outputs
1 0 0	0 1	
1 0 1	1 0	
1 1 0	1 0	
1 1 1	1 1	—— unsafe output

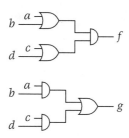

3.20

Figure

Checker for 2-out-of-4 code.

a	b	c	d	f	g
0	0	1	1	0	1
0	1	0	1	1	0
1	0	0	1	1	0
0	1	1	0	1	0
1	0	1	0	1	0
1	1	0	0	0	1
0	1	1	1	1	1
1	1	0	1	1	1
1	0	1	1	1	1
1	0	0	0	0	0
0	0	0	1	0	0
0	0	1	0	0	0
0	1	0	0	0	0
0	0	0	0	0	0
1	1	1	1	1	1

3.10

Table

Truth table for ultimate fail-safe checker for 2-out-of-4 code.

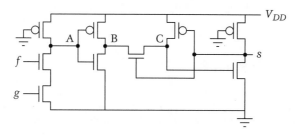

3.21

Figure

Translator circuit.

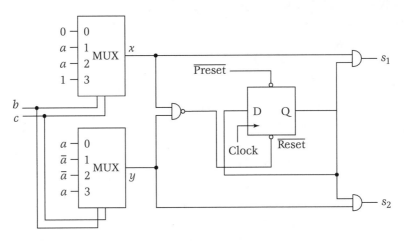

An ultimate fail-safe circuit.

The output bits x and y represent a binary number whose value equals the number of 1s at the inputs. As long as there are at most two 1s at the inputs, the output is considered to be correct. An all-0s output is assumed to be incorrect but safe, whereas an all-1s output is incorrect and also unsafe.

The implementation of the ultimate fail-safe circuit corresponding to the truth table is shown in Figure 3.22. The functional circuit is implemented using two 4-to-1 multiplexers; the D flip-flop is preset to 1. If the circuit produces the unsafe output 1 1 due to a fault in the circuit or because of all 1s at the input, the reset input of the flip-flop is activated forcing the outputs of the AND gates to be locked at 0 0, that is, $s_1 s_2 = 0\,0$ (safe output). The circuit remains at this safe output state until the flip-flop is preset again, which allows the resumption of the normal operation of the circuit. Obviously, the additional circuit incorporated to guarantee ultimate fail-safe operation must be tested off-line to eliminate any fault that may exist.

3.11   REFERENCES

[3.1]   Anderson, D. A., and G. Metze. "Design of totally self checking check circuits for *m*-out-of-*n*-codes." *IEEE Trans. Computers*, C-22 (March 1973): 263–269.

[3.2]   Marouf, M. A., and A. D. Friedman. "Efficient design of self-checking checker of any *m*-out-of-*n*-codes." *IEEE Trans. Computers*, C-27 (June 1978): 482–490.

[3.3]   Smith, J. E., and G. Metze. "Strongly fault secure logic networks." *IEEE Trans. Computers,* C-27 (June 1978): 491–499.

[3.4]   Nanya, T., and T. Kawamura. "Error secure/propagating concept and its application to the design of strongly fault-secure processors." *IEEE Trans. Computers,* C-37 (January 1988): 14–24.

[3.5]   Anderson, D. A. "Design of self checking networks using coding techniques." Ph.D. dissertation, Univ. of Illinois, Urbana, 1971.

[3.6]   Nicolaidis, M., I. Jansch, and B. Courtois. "Strongly code disjoint checkers." *Proc. 14th Int. Symp. on Fault-Tolerant Computing,* 1984, 16–21.

[3.7]   Pagey, S., S. D. Sherlekar, and G. Venkatesh. "A methodology for the design of SFS/SCD circuits for a class of unordered codes." *Jour. Electronic Testing: Theory and Applications* 2 (1991): 261–277.

[3.8]   Bartlett, K., R. K. Brayton, G. D. Hatchtel, R. M. Jacoby, C. R. Morrison, and R. L. Ruddel. "Multilevel logic minimization using implicit don't cares." *IEEE Trans. Computer-Aided Design* 7 (June 1988): 723–740.

[3.9]   Busaba, F. Y., and P. K. Lala. "Self checking combinational circuit design for single and unidirectional multibit error." *Jour. Electronic Testing: Theory and Applications* 5 (1994): 19–28.

[3.10]   Sellers Jr., F. F., M. Hsiao, and L. W. Bearnson. *Error Detecting Logic for Digital Computers.* McGraw-Hill, 1968.

[3.11]   Gupta, S. K., and D. K. Pradhan. "Can concurrent checkers help BIST?" *Proc. Int. Test Conf.,* 1992, pp. 140–150.

[3.12]   Saposhnikov VI, V., A. Dmitriev, M. Goessel, V. V. Saposhnikov. "Self-dual parity checking—a new method for on-line testing." *Proc. IEEE 14th VLSI Test Symposium,* 1996, 162–168.

[3.13]   Biswas, N. N. *Logic Design Theory.* Prentice Hall, 1993.

[3.14]   Sayers, I. L., and D. J. Kinniment. "Low-cost residue codes and their application to self-checking VLSI systems." *Proc. IEE* 132, no. 4 (July 1985): 197–202.

[3.15]   Ostapko, D. L., and S. J. Hong. "Fault analysis and test generation for programmable logic arrays." *IEEE Trans. Computers,* C-28 (September 1979): 617–626.

[3.16]   Cha, C. W. "A testing strategy for PLAs." *Proc. 15th Design Automation Conf.,* 1978, 326–334.

[3.17]   Smith, J. E. "Detection of faults in programmable logic arrays." *IEEE Trans. Computers,* C-28 (November 1979): 845–853.

[3.18]   Wang, S. L., and A. Avizienis. "The design of totally self-checking circuits using programmable logic arrays." *Proc. 9th Int. Symp. on Fault-Tolerant Computing,* 1979, 173–180.

[3.19]   Mak, G. P., J. A. Abraham, and E. S. Davidson. "The design of PLAs with concurrent error detection." *Proc. 12th Int. Symp. on Fault-Tolerant Computing,* 1982, 303–310.

[3.20] Arevalo, Z., and J. G. Bredson. "A method to simplify a Boolean function into a near minimal sum-of-products for programmable logic arrays." *IEEE Trans. Computers,* C-27 (November 1978): 1028–1039.

[3.21] Tao, D. L., P. K. Lala, and C. R. P. Hartmann. "A concurrent testing strategy for PLAs." *Proc. 1986 Int. Test Conf.,* 1986, 705–709.

[3.22] Mine, H., and Y. Koga. "Basic properties and a construction method for fail-safe logic systems." *IEEE Trans. Electronic Computers* 16 (June 1967): 282–289.

[3.23] Nicolaidis, M. "Fail-safe interfaces for VLSI: Theoretical foundations and implementations." *IEEE Trans. Computers* 47 (January 1998): 62–77.

4 | Self-Checking Checkers

Considerable work has been done in recent years on the design of totally self-checking checkers for various types of error detecting codes. A totally self-checking checker must have two outputs and, hence, four output combinations (Figure 4.1). Two of these output combinations—for example, (0 1, 1 0)—are considered as valid. A nonvalid checker output, 0 0 or 1 1, indicates either a non-code word at the input of the checker or a fault in the checker itself. The reason a checker needs two outputs is because if there is only one output and the normal output value is, for instance, 1, then a stuck-at-1 fault at the output cannot be detected during normal operation. Also, output combinations (0 0, 1 1) are not selected as valid because a unidirectional multibit error may change 0 0 to 1 1 or vice versa.

It should be pointed out that a checker does not need to be fault-secure because one is interested only in whether the output of the checker circuit is a code word or not—that is, 0 1 or 1 0. If the output is a valid code word, it is not important whether 0 1 has been changed to 1 0 or vice versa because as long as the checker is self-testing, the output of the checker will be 0 0 or 1 1 in the presence of a fault [4.1].

4.1   THE TWO-RAIL CHECKER

The two-rail checker has two groups of inputs $(x_1, x_2, \ldots, x_n)$ and $(y_1, y_2, \ldots, y_n)$ and two outputs f and g. The signals observed on the outputs should always

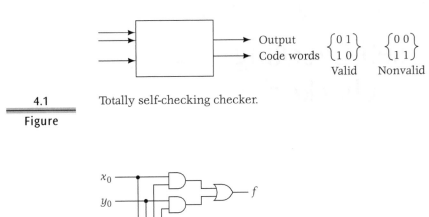

4.1

Figure

Totally self-checking checker.

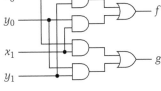

4.2

Figure

Totally self-checking two-rail checker.

be complementary, that is, a 1-out-of-2 code if and only if every pair x_j, y_j is also complementary for all $j (1 \leq j \leq n)$. This technique can be understood by looking at the circuit of Figure 4.2, where $y_i = \bar{x}_i$. In a nonerror situation when $x_0 x_1 = 1\,1$, $y_0 y_1 = 0\,0$; the result of this is $f = 0$, $g = 1$. Now consider a situation where due to a fault $y_0 y_1 = 1\,0$. The circuit of Figure 4.2 will then produce $f = g = 1$, a non-code output thus giving an error indication. In fact the circuit is totally self-checking for all single and unidirectional multiple faults.

Although a two-rail checker for an arbitrary number of input pairs may be designed using two-level AND-OR logic, it is more efficiently realized as a tree by interconnecting the checker modules with two input pairs [4.2]. In general a multilevel tree realization for a checker with m input pairs, formed by interconnecting checker modules with x input pairs, requires $\lceil (m - 1)/(x - 1) \rceil$ modules and $\lceil \log_2 m \rceil$ module levels. For example, Figure 4.3 shows a two rail checker with $m = 6$, formed by interconnecting checker modules with $x = 2$ input pairs (Figure 4.2). Anderson and Metze [4.2] have shown that such a tree can be tested with 2^x input combinations, that is, all possible input code words to a module. Thus, the test set for the circuit of Figure 4.3 is the normal input set $\{0\,1\,0\,1, 1\,0\,1\,0, 0\,1\,1\,0, 1\,0\,0\,1\}$ of the circuit of Figure 4.2.

A technique for designing an area-time efficient two-rail checker in static CMOS technology has been proposed by Lo [4.3]. An n-pair two-rail checker is constructed from a two-pair two-rail checker as shown in Figure 4.4. The two

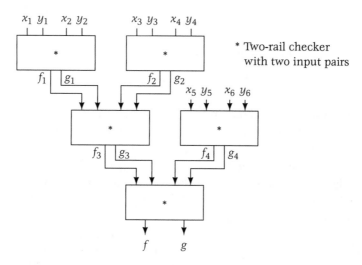

4.3

Figure

Totally self-checking two-rail checker with six input pairs.

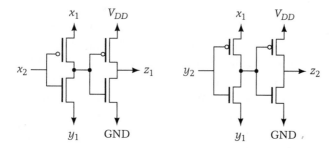

4.4

Figure

Two-rail checker (from [4.3], courtesy of IEEE, ©1993).

inputs to the cell are (x_1, y_1) and (x_2, y_2) where $x_1 = \overline{y}_1$ and $x_2 = \overline{y}_2$. The outputs of the cell are also in two-rail form, $z_1 = \overline{z}_2$. The checker is totally self-checking with respect to the following set of faults:

+ stuck-at-0/1 at input and output lines
+ bridging between input lines
+ breaks in input lines
+ bridging in source-drain, gate-source, and gate-drain of each transistor
+ stuck-on and stuck-open transistors

4.2 TOTALLY SELF-CHECKING CHECKERS FOR *m*-OUT-OF-*n* CODES

An *m*-out-of-*n* code is one in which all valid code words have exactly *m* 1s and $(n - m)$ 0s. As discussed in Chapter 2, such codes are "nonseparable" codes, because the information is embedded in the code word with the redundancy. If the information is required in another form elsewhere in the system, a code translator is needed to convert the coded information into the desired format.

The *m*-out-of-*n* checker consists of two independent subcircuits, each subcircuit having a single output. For normal *m*-out-of-*n* code inputs, the checker output is 0 1 or 1 0. If the number of 1s at the checker input is greater or less than *m* (i.e., invalid code inputs), then the output is 1 1 or 0 0, respectively. In order for the *m*-out-of-*n* checker to be self-testing, the code words must contain the same number of 1s and 0s, that is, $n = 2m$. Codes of this type are known as *k*-out-of-2*k* codes.

The *k*-out-of-2*k* checker is fault-secure for single faults because it has two subcircuits; a single fault can affect the output of only one of the subcircuits. Hence the checker is totally self-checking for all single faults. If the checker is realized with AND-OR logic, it is also totally self-checking for unidirectional multiple faults because it contains no inverters. Thus, a unidirectional multiple fault in the checker can cause only a faulty 1 or faulty 0 at the output but not a faulty 1 or 0 at the same time. The number of tests required to diagnose all single and unidirectional multiple faults in the checker is 2^k.

To design the *k*-out-of-2*k* checker, the 2*k* bits are partitioned into two disjoint subsets $A(x_1, x_2, \ldots, x_k)$ and $B(x_{k+1}, x_{k+2}, \ldots, x_{2k})$. If the two outputs of the checker are designated as Z_1 and Z_2, then

$$Z_1 = \sum_{i=1}^{k} T(k_A \geq i) \cdot T(k_B \geq k - i) \quad (i = 1, 3, 5, \ldots, \text{ an odd number})$$

$$Z_2 = \sum_{i=0}^{k} T(k_A \geq i) \cdot T(k_B \geq k - i) \quad (i = 0, 2, 4, \ldots, \text{ an even number})$$

where k_A and k_B are the number of 1s occurring in subsets *A* and *B*, respectively. $T(k_A \geq i)$ represents the function that has the value 1 if and only if the number of 1s in subset *A* is greater than or equal to the value *i*; similarly,

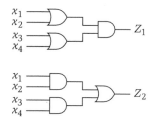

4.5

Figure

Logic circuit for the 2-out-of-4 checker.

$T(k_B \geq k - i)$ has the value 1 if and only if the number of 1s in subset B is greater than or equal to the value $(k - i)$.

As an example of this checker, the design of a totally self-checking 2-out-of-4 checker is described. In this example $k = 2$ and $A = (x_1, x_2)$ and $B = (x_3, x_4)$. The two output functions become

$$Z_1 = T(k_A \geq 1) \cdot T(k_B \geq 1)$$
$$= (x_1 + x_2) \cdot (x_3 + x_4)$$
$$Z_2 = T(k_A \geq 0) \cdot T(k_B \geq 2) + T(k_A \geq 2) \cdot T(k_B \geq 0)$$
$$= 1 \cdot T(k_B \geq 2) + T(k_A \geq 2) \cdot 1$$
$$= x_3 x_4 + x_1 x_2$$

The circuit diagram for the 2-out-of-4 checker is shown in Figure 4.5.

A general m-out-of-n checker, where $n \neq 2k$, can be realized by translating the given code to a 1-out-of-$\binom{n}{m}$ code, which is converted to a k-out-of-$2k$ code

$$\left[2^k \leq \binom{n}{m} \leq \binom{2k}{k} \right]$$

via a totally self-checking translator. As an example, we describe the design of a 2-out-of-5 code checker. The design consists of changing the 2-out-of-5 code into a 1-out-of-10 code, which is then translated into a 3-out-of-6 code. Figure 4.6 shows the circuit for decoding the 2-out-of-5 code into the 1-out-of-10 code. The decoder is totally self-checking and consists of ten two-input AND gates; each AND gate recognizes only one code word input. Table 4.1 shows the decoded output.

The 1-out-of-10 code is translated into a 3-out-of-6 code (Table 4.2) via a single level of OR gate as shown in Figure 4.7.

A systematic procedure for designing two-level totally self-checking checkers for m-out-of-$(2m + 1)$, $m > 1$, codes has been presented by Reddy [4.4].

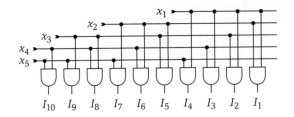

4.6

Figure

A 2-out-of-5 to 1-out-of-10 decoder.

2-out-of-5 code					1-out-of-10 code									
x_1	x_2	x_3	x_4	x_5	I_1	I_2	I_3	I_4	I_5	I_6	I_7	I_8	I_9	I_{10}
1	1	0	0	0	1	0	0	0	0	0	0	0	0	0
1	0	1	0	0	0	1	0	0	0	0	0	0	0	0
1	0	0	1	0	0	0	1	0	0	0	0	0	0	0
1	0	0	0	1	0	0	0	1	0	0	0	0	0	0
0	1	1	0	0	0	0	0	0	1	0	0	0	0	0
0	1	0	1	0	0	0	0	0	0	1	0	0	0	0
0	1	0	0	1	0	0	0	0	0	0	1	0	0	0
0	0	1	1	0	0	0	0	0	0	0	0	1	0	0
0	0	1	0	1	0	0	0	0	0	0	0	0	1	0
0	0	0	1	1	0	0	0	0	0	0	0	0	0	1

4.1

Table

2-out-of-5 to 1-out-of-10 code conversion.

1-out-of-10 code										3-out-of-6 code					
I_1	I_2	I_3	I_4	I_5	I_6	I_7	I_8	I_9	I_{10}	Z_1	Z_2	Z_3	Z_4	Z_5	Z_6
1	0	0	0	0	0	0	0	0	0	1	1	1	0	0	0
0	1	0	0	0	0	0	0	0	0	1	1	0	0	0	1
0	0	1	0	0	0	0	0	0	0	1	0	1	0	1	0
0	0	0	1	0	0	0	0	0	0	1	0	0	0	1	1
0	0	0	0	1	0	0	0	0	0	0	1	1	1	0	0
0	0	0	0	0	1	0	0	0	0	0	1	0	1	0	1
0	0	0	0	0	0	1	0	0	0	0	0	1	1	1	0
0	0	0	0	0	0	0	1	0	0	0	0	0	1	1	1
0	0	0	0	0	0	0	0	1	0	0	1	0	0	1	1
0	0	0	0	0	0	0	0	0	1	1	0	1	1	0	0

4.2

Table

1-out-of-10 to 3-out-of-6 code conversion.

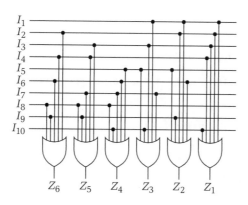

4.7

Figure

A 1-out-of-10 to 3-out-of-6 translator.

Such checkers are basically composed of two subcheckers, an m-out-of-$(2m + 1)$ to 1-out-of-$^{(2m+1)}C_m$ code translator, and a 1-out-of-$^{(2m+1)}C_m$ to 1-out-of-2 code translator. The first translator is realized by using $^{(2m+1)}C_m$ number of m-input AND gates. The second translator consists of only two OR gates, with a total of $^{(2m+1)}C_m$ inputs. The output of each AND gate is connected to an input of a single OR gate. Hence the set of AND gates has to be partitioned into two disjoint subsets.

The partitioning of the set of AND gates into two subgroups, Group I and Group II, is shown in Figure 4.8. A code word in a partitioned set is represented in two different ways. In the first representation, indications of the positions of 1s in a code word are given. For example, 0 0 1 1 1 is represented by 345, and 1 1 0 0 0 0 1 is represented by 127. In the second representation a code word is assumed to be divided into three regions A, B, C, with both A and C containing m bits and B containing only 1 bit. A code word is then represented by the 3-tuple $(a|b|c)$, where a, b, and c are the number of 1s in the A, B, and C regions, respectively. For example, 0 0 1 1 1 is represented by (0|1|2), and 1 1 0 0 0 0 1 is represented by (2|0|1). The second type of representation is used in Figure 4.8 except for $A^\dagger$, for which the first type of representation is used.

As an example, we consider the design of a totally self-checking 2-out-of-5 code checker circuit. Table 4.3 shows the truth table for the first translator circuit. This can be designed using the two-input AND gates.

The set of AND gates is partitioned into two groups as shown in Figure 4.9. Each group represents the subset of AND gates that feed an OR gate. The complete circuit for the 2-out-of-5 checker is shown in Figure 4.10.

Group I

$$A_i = \{i|0|(m-i), i \text{ even and } 2 \le i \le m\}$$

$$A^\dagger = \{i(m+i_1)(m+i_2)\ldots(m+i_{m-1}), 1 \le i \le m, 2 \le i_j \le (m+1), \text{ and } i_j - 1 \ne i\}$$

and

$$B_i = \{i|1|(m-i-1), i \text{ even and } 0 \le i \le m-1\}$$

Group II

$$A_i = \{i|0|(m-i), i \text{ odd and } 3 \le i \le m\}$$

$$A^* = \{\text{all elements of } (1|0|(m-1)) \text{ not included in } A^\dagger\}$$

$$A_0 = (0|0|m)$$

and

$$B_j = \{j|1|(m-1-j), j \text{ odd and } 1 \le j \le m-1\}$$

4.8 Figure	Partitioning of a set of AND gates (from [4.4], courtesy of IEEE, ©1974).

2-out-of-5 code					1-out-of-10 code									
x_1	x_2	x_3	x_4	x_5	I_1	I_2	I_3	I_4	I_5	I_6	I_7	I_8	I_9	I_{10}
1	1	0	0	0	1	0	0	0	0	0	0	0	0	0
1	0	1	0	0	0	1	0	0	0	0	0	0	0	0
1	0	0	1	0	0	0	1	0	0	0	0	0	0	0
1	0	0	0	1	0	0	0	1	0	0	0	0	0	0
0	1	1	0	0	0	0	0	0	1	0	0	0	0	0
0	1	0	1	0	0	0	0	0	0	1	0	0	0	0
0	1	0	0	1	0	0	0	0	0	0	1	0	0	0
0	0	1	1	0	0	0	0	0	0	0	0	1	0	0
0	0	1	0	1	0	0	0	0	0	0	0	0	1	0
0	0	0	1	1	0	0	0	0	0	0	0	0	0	1

4.3 Table	Truth table for 2-out-of-5 code to 1-out-of-10 code.

Group I	Group II
$A_2 = \{1\,1\,0\,0\,0\}$	$A^* = \{0\,1\,0\,0\,1,\ 1\,0\,0\,1\,0\}$
$A^\dagger = \{15, 24\} = \{1\,0\,0\,0\,1,\ 0\,1\,0\,1\,0\}$	$A_0 = \{0\,0\,0\,1\,1\}$
$B_0 = \{0\,0\,1\,0\,1,\ 0\,0\,1\,1\,0\}$	$B_1 = \{1\,0\,1\,0\,0,\ 0\,1\,1\,0\,0\}$

4.9

Figure

Partitioning of the set of AND gates.

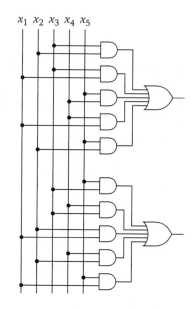

4.10

Figure

Totally self-checking checker for 2-out-of-5 code (from [4.4], courtesy of IEEE, ©1974).

Reddy [4.4] has also developed a procedure for cellular realization of totally self-checking checkers for k-out-of-$2k$ codes. This method uses the same technique as Anderson's method in deriving the logical expressions for the two outputs. For example, when $k = 4$, the expressions for the outputs are

$$Z_1 = \sum_{i=0}^{4} T(k_A \geq i) \cdot T(k_b \geq 4 - i) \quad (i = 1, 3)$$

$$Z_2 = \sum_{i=0}^{4} T(k_A \geq i) \cdot T(k_b \geq 4 - i) \quad (i = 0, 2, 4)$$

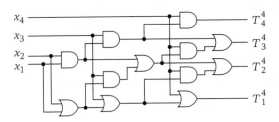

4.11

Figure

T^4 array.

Reddy and Wilson [4.5] have shown that functions $T(k_A \geq i)$ and $T(k_B \geq j)$ can be realized using two-dimensional cellular logic networks. An n-variable cellular array, T^n, can realize all the nT_p^n functions of the n variables (a T_p^n function takes the value 1 if and only if p or more of the n variables are 1). The T_n array requires $(n \cdot (n + 1)/2 - 1)$ cells containing $(n \cdot (n - 1)/2)$ AND gates and $(n \cdot (n + 1)/2)$ OR gates; each gate has a fan-in of two and a maximum fan-out of two. For example, a T^4 array can realize the functions T_1^4, T_2^4, T_3^4, and T_4^4 as shown in Figure 4.11.

The output expressions Z_1 and Z_2 for a k-out-of-$2k$ checker can be expressed in the following form:

$$Z_1 = T_{1A}^4 \cdot T_{3B}^4 + T_{3A}^4 \cdot T_{1B}^4$$

$$Z_2 = T_{4B}^4 \cdot T_{2A}^4 + T_{2B}^4 \cdot T_{4A}^4$$

Thus, Z_1 and Z_2 can be realized using two T^4 arrays, three AND gates, and two OR gates, as shown in Figure 4.12.

It is shown in reference [4.5] that a checker designed with cellular array requires only $2k$ tests to detect all stuck-at faults. The test set is given below.

$$
\text{Test set for } k\text{-out-of-}2k \text{ checkers} =
\begin{vmatrix}
1 & 1 & 1 & \ldots & \ldots & 1 & 0 & 0 & \ldots & \ldots & \ldots & 0 \\
0 & 1 & 1 & \ldots & \ldots & 1 & 1 & 0 & \ldots & \ldots & \ldots & 0 \\
0 & 0 & 1 & \ldots & \ldots & 1 & 1 & 0 & \ldots & \ldots & \ldots & 0 \\
0 & 0 & 0 & \ldots & \ldots & 1 & 1 & 1 & 1 & \ldots & \ldots & 0 \\
1 & 1 & \ldots & \ldots & 1 & 0 & 0 & 0 & \ldots & \ldots & 0 & 1 \\
1 & 1 & \ldots & 1 & 1 & 0 & 0 & 0 & \ldots & 0 & 1 & 1 \\
1 & 0 & \ldots & \ldots & \ldots & 0 & 0 & 1 & \ldots & \ldots & 1 & 1 \\
0 & 0 & \ldots & \ldots & \ldots & 0 & 1 & 1 & \ldots & \ldots & \ldots & 1
\end{vmatrix}
$$

$$\longleftarrow k \longrightarrow \qquad \longleftarrow k \longrightarrow$$

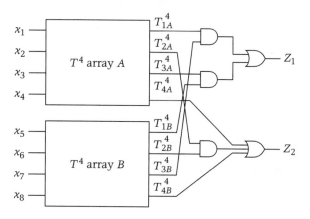

4.12

Figure

Cellular realization of a 4-out-of-8 checker.

Therefore, the test set for the 4-out-of-8 checker is

$$\begin{vmatrix} 1\,1\,1\,1 & 0\,0\,0\,0 \\ 0\,1\,1\,1 & 1\,0\,0\,0 \\ 0\,0\,1\,1 & 1\,1\,0\,0 \\ 0\,0\,0\,1 & 1\,1\,1\,0 \\ 1\,1\,1\,0 & 0\,0\,0\,1 \\ 1\,1\,0\,0 & 0\,0\,1\,1 \\ 1\,0\,0\,0 & 0\,1\,1\,1 \\ 0\,0\,0\,0 & 1\,1\,1\,1 \end{vmatrix}$$

Smith also proposed a method for cellular realization of two-output k-out-of-$2k$ totally self-checking checkers [4.6]. Figure 4.13 illustrates the construction of k-out-of-$2k$ checkers; such checkers can be tested by $2k$ code words resulting from the circular shift of k 1s followed by k 0s.

As an example, the cellular realization of 4-out-of-8 codes is shown in Figure 4.14(a); the test words for the checker are given in Figure 4.14(b). In general, k-out-of-$2k$ checkers designed by Smith's approach will be faster and require fewer gates than that of Reddy's [4.4]. For example, the 4-out-of-8 checker, if designed using Reddy's procedure, will need $7(= 2k - 1)$ levels of logic and $29(= 2k^2 - k + 1)$ gates, whereas Smith's technique requires $4(= k)$ levels of logic and $26(= 2k^2 - 2k + 2)$ gates as shown in Figure 4.14(a). The number of test words derived by each method will be the same, for example, $8(= 2k)$ in the case of 4-out-of-8 checkers [4.6].

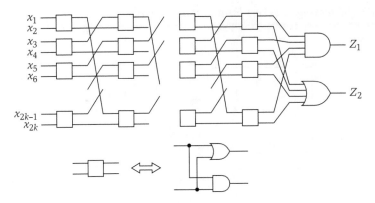

4.13

Figure

A totally self-checking checker for k-out-of-$2k$ codes (from [4.5], *courtesy of IEEE, ©1974*).

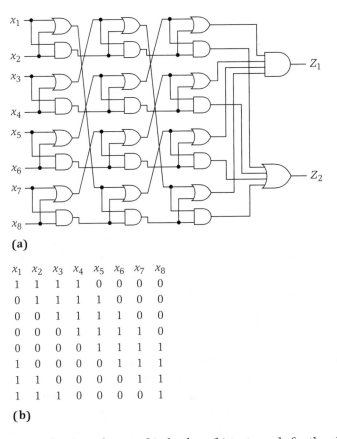

(a)

x_1	x_2	x_3	x_4	x_5	x_6	x_7	x_8
1	1	1	1	0	0	0	0
0	1	1	1	1	0	0	0
0	0	1	1	1	1	0	0
0	0	0	1	1	1	1	0
0	0	0	0	1	1	1	1
1	0	0	0	0	1	1	1
1	1	0	0	0	0	1	1
1	1	1	0	0	0	0	1

(b)

4.14

Figure

(a) Realization of 4-out-of-8 checker; (b) test words for the checker.

Marouf and Friedman [4.7] have also presented a procedure for designing totally self-checking checkers for any arbitrary m-out-of-n code, $m \geq 2$. These checkers are also totally self-checking for single and unidirectional multiple faults. Let us first describe the procedure for the case when $2m + 2 \leq n \leq 4m$. It consists of the following steps:

1. Partition the n bits into sets A and B such that the number of bits in set A is $n_a = \lfloor n/2 \rfloor$ and the number of bits in set B is $n_b = n - n_a = \lceil n/2 \rceil$. Define two functions f_1 and f_2 as

$$f_1 = \sum_{i=1}^{m-1} T(a \geq i) T(b \geq m - i) \quad (i \text{ odd})$$

$$f_2 = \sum_{i=1}^{m-1} T(a \geq i) T(b \geq m - i) \quad (i \text{ even})$$

where a and b are the number of 1s occurring in set A and B, respectively.

2. Partition set A into two subsets A_1 and A_2 such that the number of bits in subsets A_1 and A_2 are $n_{a_1} = \lfloor n_a/2 \rfloor$ and $n_{a_2} = n_a - n_{a_1} = \lceil n_a/2 \rceil$, respectively. Define f_3 and f_4 as

$$f_4 = \sum_{i=m-n_{a_2}}^{n_{a_1}} T(a_1 \geq i) T(a_2 \geq m - i) \quad (i \text{ odd})$$

$$f_4 = \sum_{i=m-n_{a_2}}^{n_{a_1}} T(a_1 \geq i) T(a_1 \geq m - i) \quad (i \text{ even})$$

3. Partition set B into two subsets B_1 and B_2 such that the number of bits in subsets B_1 and B_2 are $n_{b_1} = \lfloor n_b/2 \rfloor$ and $n_{b_2} = n_b - n_{b_1} = \lceil n_b/2 \rceil$. Define f_5 and f_6 as

$$f_5 = \sum_{i=m-n_{b_2}}^{n_{b_1}} T(b_1 \geq i) T(b_2 \geq m - i) \quad (i \text{ odd})$$

$$f_6 = \sum_{i=m-n_{b_2}}^{n_{b_1}} T(b_1 \geq i) T(b_2 \geq m - i) \quad (i \text{ even})$$

4. Design a circuit with n inputs and $f_i (i = 1, 6)$ outputs; realize every f_i in sum-of-products form. Call this circuit C_1, which produces a 1-out-of-6 code on its outputs.

5. Feed the outputs of circuit C_1 to the inputs of circuit C_2, which is a totally self-checking 1-out-of-6 to 2-out-of-4 translator.

6. Feed the outputs of circuit C_2 to the inputs of circuit C_3, which is a totally self-checking checker for a 2-out-of-4 code.

The combined circuit C is a totally self-checking checker. All single and unidirectional faults in C can be detected by using T code words, where

$$T = \sum_{i=1}^{m-1} \max[C_i^{n_a}, C_{m-i}^{n_b}] + \sum_{i=m-n_{a_2}}^{n_{a_1}} \max[C_i^{n_{a_1}}, C_{m-i}^{n_{a_2}}]$$

$$+ \sum_{i=m-n_{b_2}}^{n_{b_1}} \max[C_i^{n_{b_1}}, C_{m-i}^{n_{b_2}}]$$

Let us apply the above procedure to design a checker for a 2-out-of-6 code defined by the set of inputs

$$\{x_1, x_2, x_3, x_4, x_5, x_6\}$$

Then

$$A = \{x_1, x_2, x_3\} \qquad B = \{x_4, x_5, x_6\} \qquad n_a = n_b = 3$$

$$f_1 = \sum_{i=1}^{2-1} T(a \geq i)T(b \geq 2 - i)$$

$$= (x_1 + x_2 + x_3)(x_4 + x_5 + x_6)$$

$$f_2 = \sum_{i=1}^{2-1} T(a \geq i)T(b \geq 2 - i)$$

$$= 0$$

$$A_1 = \{x_1\} \qquad A_2 = \{x_2, x_3\} \qquad n_{a_1} = 1 \qquad n_{a_2} = 2$$

$$f_3 = \sum_{i=0}^{1} T(a_1 \geq i)T(a_2 \geq 2 - i)$$

$$= x_1(x_2 + x_3)$$

$$f_4 = \sum_{i=0}^{1} T(a_1 \geq i)T(a_2 \geq 2 - i)$$

$$= x_2 x_3$$

$$B_1 = \{x_4\} \qquad B_2 = \{x_5, x_6\} \qquad n_{b_1} = 1 \qquad n_{b_2} = 2$$

$$f_5 = \sum_{i=0}^{1} T(b_1 \geq i)T(b_2 \geq 2 - i)$$

$$= x_4(x_5 + x_6)$$

$$f_6 = \sum_{i=0}^{1} T(b_1 \geq i)T(b_2 \geq 2 - i)$$

$$= x_5 x_6$$

Since $f_2 = 0$, C_1 will have only five outputs. Thus, C_2 will be a 1-out-of-5 to 2-out-of-4 code translator. The circuit C for the checker is shown in Figure 4.15. The number of code words required to test the checker of Figure 4.15 is

$$T = \sum_{i=1}^{1} \max[C_i^3, C_{2-i}^3] + 2 \sum_{i=0}^{1} \max[C_i^1, C_{2-i}^2]$$

$$= 3 + 2(1 + 2) = 9$$

The checker design procedure for the case $n = 2m + 1$, $m \geq 2$, is as follows:

1. Partition the n bits into two sets A and B such that the number of bits in A and B are $n_a = m$ and $n_b = m + 1$, respectively.

2. Partition set B into two subsets B_1 and B_2 such that the number of bits in B_1 and B_2 are $n_{b_1} = \lfloor n_b/2 \rfloor$ and $n_{b_2} = n_b - n_{b_1} = \lceil n_b/2 \rceil$.

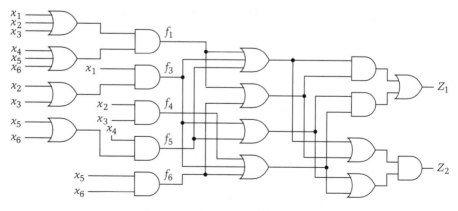

4.15

Figure

Totally self-checking checker for a 2-out-of-6 code.

3. Define the four functions f_1, f_2, f_3, and f_4 as follows:

$$f_1 = \sum_{i=1}^{m} T(a \geq i)T(b \geq m - i) \quad (i \text{ odd})$$

$$f_2 = \sum_{i=1}^{m} T(a \geq i)T(b \geq m - i) \quad (i \text{ even})$$

$$f_3 = \sum_{i=m-n_{b_2}}^{n_{b_2}} T(b_1 \geq i)T(b_2 \geq m - i) \quad (i \text{ odd})$$

$$f_4 = \sum_{i=m-n_{b_2}}^{n_{b_2}} T(b_1 \geq i)T(b_2 \geq m - i) \quad (i \text{ even})$$

4. Design a circuit with n inputs and $f_i (i = 1, 4)$ outputs. Call this circuit C_1. The output of C_1 feeds the input of circuit C_2, which is a totally self-checking 1-out-of-4 to 2-out-of-4 translator. The output of C_2 feeds the input of circuit C_3, which is a totally self-checking checker for a 2-out-of-4 code.

As an example let us design the checker for a 2-out-of-5 code defined by the set of inputs

$$\{x_1, x_2, x_3, x_4, x_5\}$$

Then

$$A = \{x_1, x_2\} \quad B = \{x_3, x_4, x_5\} \quad n_a = 2 \quad n_b = 3$$
$$B_1 = \{x_3\} \quad B_2 = \{x_4, x_5\} \quad n_{b_1} = 1 \quad n_{b_2} = 2$$

$$f_1 = \sum_{i=1}^{2} T(a \geq i)T(b \geq 2 - i) = (x_1 + x_2)(x_3 + x_4 + x_5)$$

$$f_2 = \sum_{i=1}^{2} T(a \geq i)T(b \geq 2 - i) = x_1 x_2$$

$$f_3 = \sum_{i=0}^{1} T(b_1 \geq i)T(b_2 \geq 2 - i) = x_3(x_4 + x_5)$$

$$f_4 = \sum_{i=0}^{1} T(b_1 \geq i)T(b_2 \geq 2 - i) = x_4 x_5)$$

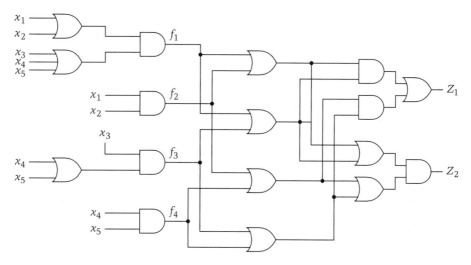

Totally self-checking checker for a 2-out-of-5 code.

The combined circuit for the given m-out-of-$(2m + 1)$ code is shown in Figure 4.16. All single and unidirectional faults in the checker can be detected by using T' code words, where

$$T' = \sum_{i=1}^{m} C_{m-i}^{m+1} + \sum_{i=m-n_{b_2}}^{n_{b_1}} \max[C_i^{n_{b_1}}, C_{m-i}^{n_{b_2}}]$$

For the circuit shown in Figure 4.16,

$$T' = \sum_{i=1}^{2} C_{2-i}^{3} + \sum_{i=0}^{1} \max[C_i^1, C_{2-i}^2]$$

$$= (3 + 1) + (1 + 2) = 7 \text{ code words}$$

4.2.1 Pass Transistor-Based Checker Design for a Subset of m-out-of-2m Codes

A totally self-checking checker for a selected subset of m-out-of-2m code can be designed by using a totally self-checking checker cell for the 2-out-of-4 code

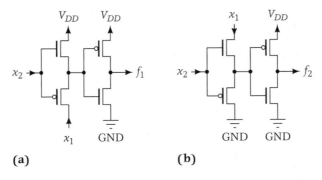

4.17

Figure

Transistor-level implementation. (a) NOR; (b) NAND.

[4.8]. The Boolean expressions corresponding to a totally self-checking checker for 2-out-of-4 code are

$$z_1 = (x_1 + x_2)(x_3 + x_4) = \overline{\overline{(x_1 + x_2)} + \overline{(x_3 + x_4)}}$$

$$z_2 = x_1 x_2 + x_3 x_4 = \overline{\overline{x_1 x_2} \cdot \overline{x_3 x_4}}$$

where x_1, x_2, x_3, and x_4 are the inputs and z_1 and z_2 are the outputs. These expressions can be implemented using pass transistor-based structures of NOR and NAND functions. Figure 4.17 shows the implementations of the NOR and the NAND functions.

The 2-out-of-4 checker cell is implemented using three NAND and three NOR gates as shown in Figure 4.18; each of these gates is shown as a black box in the design. The outputs of the 2-out-of-4 checker for non-code word and valid code word inputs are shown in Table 4.4(a) and (b), respectively. The checker is totally self-checking for stuck-at, stuck-open, breaks, and bridging faults.

The 2-out-of-4 checker cell can be used for constructing checkers for m-out-of-$2m$ ($m = 3, 4, 5, 6$) code. The $2m$ bits are partitioned into two blocks A_1 and B_1:

$$A_I = \{x_1, x_2, \ldots, x_{m+1}\}, \quad B_I = \{x_{m+2}, x_{m+3}, \ldots, x_{2m}\} \text{ for } m = 3, 5$$

$$A_I = \{x_1, x_2, \ldots, x_m\}, \quad B_I = \{x_{m+1}, x_{m+2}, \ldots, x_{2m}\} \text{ for } m = 4, 6$$

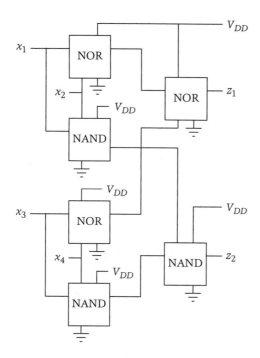

A 2-out-of-4 checker.

(a)	Non-code word $x_1x_2x_3x_4$	Output z_1z_2	(b)	Code word $x_1x_2x_3x_4$	Output z_1z_2
	0 0 0 0	0 0		0 0 1 1	0 1
	0 0 0 1	0 0		0 1 0 1	1 0
	0 0 1 0	0 0		0 1 1 0	1 0
	0 1 0 0	0 0		1 0 0 1	1 0
	1 0 0 0	0 0		1 0 1 0	1 0
	0 1 1 1	1 1		1 1 0 0	0 1
	1 0 1 1	1 1			
	1 1 0 1	1 1			
	1 1 1 0	1 1			
	1 1 1 1	1 1			

(a) Outputs for non-code words; (b) outputs for code words.

The elements of the blocks of A_1 and B_1 are then rearranged for different values of m as follows:

$m = 3$ $A_I = \{(x_1, x_2)(x_3, x_4)\}$

$B_I = (x_5, x_6)$

$m = 4$ $A_I = \{(x_1, x_2)(x_3, x_4)\}$

$B_I = \{(x_5, x_6)(x_7, x_8)\}$

$m = 5$ $A_I = \{(x_1, x_2)(x_3, x_4)(x_5, x_6)\}$

$B_I = \{(x_7, x_8)(x_9, x_{10})\}$

$m = 6$ $A_I = \{(x_1, x_2, x_3)(x_4, x_5, x_6)\}$

$B_I = \{(x_7, x_8, x_9)(x_{10}, x_{11}, x_{12})\}$

Next A_{II} and B_{II} partitions for each value of m are derived as follows:

$m = 3$ $A_{II} = (x_3, x_4, x_5, x_6)$

$B_{II} = (x_1, x_2)$

$m = 4$ $A_{II} = (x_3, x_4, x_7, x_8)$

$B_{II} = (x_1, x_2, x_5, x_6)$

$m = 5$ $A_{II} = (x_3, x_4, x_5, x_6, x_9, x_{10})$

$B_{II} = (x_1, x_2, x_7, x_8)$

$m = 6$ $A_{II} = (x_4, x_5, x_6, x_{10}, x_{11}, x_{12})$

$B_{II} = (x_1, x_2, x_3, x_7, x_8, x_9)$

For $m = 5$ and 6 the elements of the blocks of A_{II} and B_{II} are rearranged as shown below:

$m = 5$ $A_{III} = (x_5, x_6, x_7, x_8, x_9, x_{10})$

$B_{III} = (x_1, x_2, x_3, x_4)$

$m = 6$ $A_{III} = (x_1, x_2, x_3, x_{10}, x_{11}, x_{12})$

$B_{III} = (x_4, x_5, x_6, x_7, x_8, x_9)$

The totally self-checking checker design for 3-out-of-6 code ($m = 3$) is shown in Figure 4.19. It is constructed by connecting the elements in A_I to a 2-out-of-4 checker, the outputs of which, together with the elements in B_I, are connected to another 2-out-of-4 code checker. The elements in A_{II} and B_{II} are connected to another pair of 2-out-of-4 checkers in the same manner. The outputs of the second-level 2-out-of-4 checkers (i.e., $(z_1 z_2)_I$ and $(z_1 z_2)_{II}$) are connected to a totally self-checking two-rail checker.

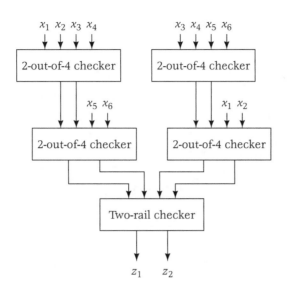

4.19   A 3-out-of-6 checker.

Figure

The checker for 4-out-of-8 ($m = 4$) can be designed in a similar manner; however, at the first level there are four 2-out-of-4 checkers (Figure 4.20).

The checkers for 5-out-of-10 code ($m = 5$) use a combination of 3-out-of-6 and 2-out-of-4 code checkers at the first level (Figure 4.21). The checker for the 6-out-of-12 code uses only a 3-out-of-6 code checker at the first level (Figure 4.22).

The above technique for designing a subset of an m-out-of-$2m$ code checker can in principle be generalized to design checkers for arbitrary m-out-of-$2m$ code. However, the distribution of inputs to form appropriate partition blocks is a challenging task.

4.2.2   Totally Self-Checking Checker for 1-out-of-n Code

One special case of an m-out-of-n code that is of much interest in computer systems is the 1-out-of-n code. Tao and Lala [4.9] have presented a new design technique for a totally self-checking 1-out-of-n checker. The checker has two subcircuits C_1 and C_2. Subcircuit C_1 receives a 1-out-of-n code and produces k-out-of-$2k$ code. Subcircuit C_2 is a totally self-checking checker for k-out-of-$2k$ code. The subcircuits are implemented using an inverter-free NOR-NOR PLA. Before describing the checker design procedure, let us consider the notations and definitions used in the procedure [4.10].

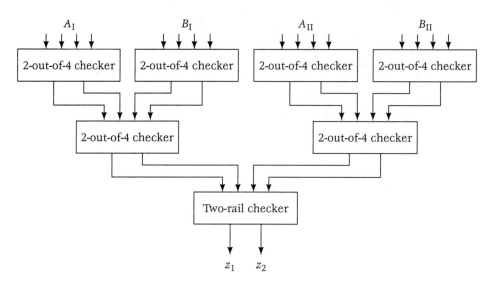

A 4-out-of-8 checker.

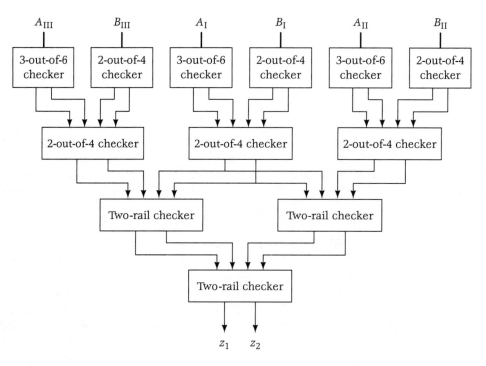

A 5-out-of-10 checker.

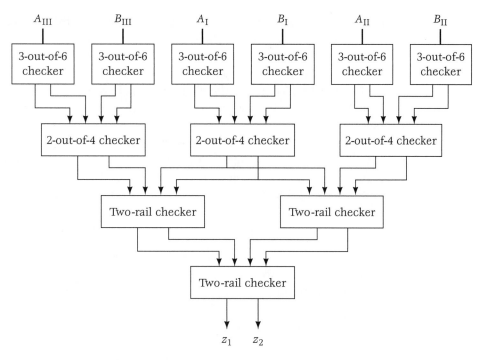

4.22

Figure

A 6-out-of-12 checker.

Let C_{2k} denote the set of all binary vectors of length $2k$ with exactly k 1s. A code word in $C_{k/2k}$ is referred to as a k-vertex. A binary vector $X = (x_1, \ldots, x_k, \ldots, x_{2k})$ covers another vector $Y = (y_1, \ldots, y_k, \ldots, y_{2k})$ if and only if $x_i \geq y_i$ for all $i = 1, 2, \ldots, 2k$. A pair of k-vertices x and y are referred to as a "1-complement pair" if and only if x and y differ only at bit positions i and $i + k$ where $1 \leq i \leq k$. For example, the k-vertices $X = 1\,0\,1\,1\,0\,1\,0\,0$ and $Y = 1\,1\,1\,1\,0\,0\,0\,0$ form a 1-complement pair.

Let $V = \{(v_1, \ldots, v_k, \ldots, v_{2k}) \mid v_i(1, 0, C), 1 \leq i \leq k\}$ be the set of all k-bit ternary vectors. Let V_s be a subset of V containing all vectors with i1s, i 0s, and $(k - 2i)C$s; C can assume a value of either 0 or 1. It should be clear that no more than $\lfloor k/2 \rfloor$ bits can be at 1 in any vector in V_s. A vector in V_s with exactly i 1s is denoted by u_i. Based on $u_i = (v_1, \ldots, v_j, \ldots, v_k)$, a subset of $C_{k/2k}$ vectors, $S(u_i)$, can be constructed. Vector x can be included in $S(u_i)$ if and only if

1. both jth and $(j + k)$th bits of x are equal to 0 if $v_j = 0$.

2. both jth and $(j + k)$th bits of x are equal to 1 if $v_j = 1$.

3. both jth and $(j + k)$th bits of x are complements to each other if $v_j = C$.

Thus, a two-rail code with $2k$ bits can be considered as the set $S(u_0)$, since the first k bits are bit-by-bit complements of the last k bits. For a $u_1 = 10C0$, the corresponding set $S(u_1)$ will be $\{10001010, 10101001\}$. Similarly, if a $u_2 = 0C1C$, then the $S(u_2) = \{00100111, 00110110, 01100011, 01110010\}$. As can be seen, bits 3 and 7 in the binary vectors in $S(u_1)$ are complementary. Also, bit 2 and bit 4 of binary vectors in $S(u_2)$ are complementary to bits 6 and 8, respectively.

A ternary vector u_{i-1} is said to C-cover another ternary vector u_i if and only if all of the bits in u_i and u_{i-1} are the same in both vectors except two positions, which are at C in vector u_{i-1} and at 1 and 0 in u_i. For example, given $u_i = 10CC$, $u_2 = 1010$, and $u_3 = 1011$, u_i C-covers u_2 but not u_3, and u_2 does not C-cover u_3.

The design of the self-checking checker for 1-out-of-n code consists of the following two steps:

1. Design a two-level totally self-checking checker for the k-out-of-$2k$ code by using n number of code words from the k-out-of-$2k$ code where

$$\binom{2k-2}{2k-4} \le s^k \le n \le \binom{2k}{k}$$

2. Construct a 1-out-of-n code to a k-out-of-$2k$ code translator circuit.

The design procedures for a k-out-of-$2k$ totally self-checking checker use code words from sets $S(u_0)$ and $S(u_i)(i \ge 1)$, unlike the method proposed in [4.11] that uses only code words from $S(u_0)$. The following five design rules take into consideration both sets $S(u_0)$ and $S(u_i)$ in designing totally self-checking checkers.

Rule 1 If any code word $x \in S(u_i)$ is used in the checker design, then the rest of the code words in $S(u_i)$ must also be used except when $i = (k-1)/2$ and n and k are odd.

Rule 2 If u_i is selected then $|S(u_i)|$—that is, the number of code words in set $S(u_i)$—should be partitioned into two sets P_0 and P_i in the following manner: $x \in S(u_i)$ belongs to $P_0(P_1)$ if and only if $(r-i)$ is odd (even), where r is the number of 1s in the first k bits of x. If k is even, then $|S(u_{k/2})| = 1$. In this case, if more than one set $S(u_{k/2})$ is used in the checker, assign them to P_0 and P_1 so that the difference between the numbers of $S(u_{k/2})$ in P_0 and P_1 is not greater than 1.

Rule 3 If u_i is selected, then at least i number of sets u_{i-1} have to be selected. These u_{i-1} sets should C-cover every 1 and every 0 in U_i.

When k is odd, $|S(u_i)| \geq 2$ for all i, $1 \leq i \leq (k-1)/2$. By applying Rule 1, only $(n-1)$ code words can be selected where n and k are odd. Thus, a single code word from $S(u_{(k-1)/2})$ has to be selected in this case, which is $(1\,0\dots1010C)$. Since $|S(1\,0\dots1010C)| = 2$, we define $S(1\,0\dots1010C+) = \{1\,0\dots1010110\dots1\,0\,0\}$ and $S(u_{(k-1)/2}) = \{1\,0\dots1010010\dots10101)\}$.

Rule 4 If $u_{(k-1)/2+}$ is selected, then Rule 3 has to be satisfied for $u_{(k-1)/2} = 1\,0\dots1010C$; in addition, the following two $u_{(k-1)/2}$ sets, $1\,0\dots101C0$ and $1\,0\dots10C01$, must also be selected.

Rules 1–4 provide sufficient conditions to select a set $S(u_i)$ for constructing the checker circuit. Rule 5 deals with the PLA checker design using the selected $S(u_i)$ sets.

Rule 5 Use a code word derived by using Rules 1–4 as a product term in a NOR-NOR inverter-free PLA. A 0 in a code word is interpreted as the presence, and a 1 as the absence, of a transistor at the intersection between $z_0(z_1)$ by placing the transistor at the intersection between $z_0(z_1)$ and the product term corresponding to code words in $P_0(P_1)$. Any two adjacent product lines in the PLA must be such that one belongs to P_0 and the other to P_1.

The PLA implemented using Rule 5 will be nonconcurrent; that is, if a selected code word appears at the input of the PLA, the corresponding product line will be activated, and its output will be within 0 1 and 1 0.

Selection of n Code Words from k-out-of-2k Code

This process consists of the selection of sets $S(u_0)$, $S(u_1)$, ..., $S(u_{k/2})$, whose union contains exactly n code words. In other words, n can be represented as follows:

$$\sum_{i=0}^{k/2} P_i \times 2^{k-2i} + m$$

where P_i, $0 \leq P_i \leq (k) \times (k-1)$, denotes the number of sets $S(u_i)$ to be selected, and m is equal to 1 if n and k are odd, otherwise it is 0. When $m = 1$, $S(u_{(k-1)/2+})$ has to be included in the $S(u_i)$ sets used for selecting n code words.

The first step in the selection process is to partition the vector space V_s into three sets: the B-set, the F-set, and the R-set. The B-set, a subset of F, and a subset of R are used to design the checker. Vectors in B satisfy Rules 1–3. These vectors also ensure that Rules 3 and 4 are satisfied when any vector

$k/2k$	B-Set	F-Set
2/4	CC	1 0
3/6	CCC, $1C0$, $C01$	$10C+$
4/8	$CCCC$, $10CC$	1 0 1 0
	$CC10$, $CC01$, $01CC$	1 0 0 1, 0 1 1 0
5/10	$CCCCC$, $10CCC$, $CCC01$,	$1010C+$
	$01CCC$, $CC10C$, $CC1C0$,	$0110C$
	$10C01$, $101C0$	$011C0$, $01C10$
6/12	$CCCCCC$, $10CCCC$, $CC10CC$,	1 0 1 0 1 0
	$CCCC10$, $1010CC$, $10CC10$,	0 1 1 0 1 0, 1 0 0 1 1 0
	$CC1010$, $0110CC$, $01CC10$,	$11C00C$
	$01CCCC$, $1001CC$, $CC0110$,	$1C10C0$, $C11C00$
	$CC01CC$, $1CC0CC$, $C1CC0C$,	
	$CC1CC0$	

4.5

Table

B-set and *F*-set for k-out-of-$2k$ ($k/2k$) codes, $2 \le k \le 6$.

in F is used in a checker circuit. The R set includes all the vectors of V_s not in B or in F. Table 4.5 shows the B-set and F-set for several k-out-of-$2k$ codes.

The following algorithm can be used to select n code words from the B, F, and R sets:

1. Set $X := B$ and $i = 1$.

2. Go to Step 3 if $n - |Y0| \ge 2^{k-2i}$, where $Y = U_{u_i \in X} S(u_i)$; otherwise go to Step 5.

3. If there is a u_i in F that represents $S(u_i)$ with 2^i code words, then $X := XU\{u_i\}$, delete the u_i from F, and go to Step 2.

4. If there is a u_i in R that represents $S(u_i)$ with 2^i code words, then $X := XU\{u_i\}$, delete the u_i from R, and go to Step 2.

5. $i := i + 1$. If $i \le k/2$, then go to Step 2, else go to Step 6.

6. If n and k are odd, then $X := XU\{u_{k/2+}\}$. Stop.

To illustrate the above procedure, let us construct the totally self-checking checker for 1-out-of-15 code. Since $k = 3$ satisfies the inequality $2^k \le n \le \binom{2k}{k}$, the 15 code words will be selected from the 3-out-of-6 code. The B-set = $\{CCC, 1C0, C01\}$ and the F-set = $\{10C+\}$ are given in Table 4.5. The R-set is $\{01C, 0C1, C10, 10C-\}$.

u_i	P_0	S_i	P_i	
CCC	011100	101010	111000	100011
	110001	000111	010101	001110
1*C*0	110100		100110	
*C*01	101001		001101	
01*C*	011010		010011	
10*C*+	101100			

Fifteen code words selected from 3-out-of-6 code.

1. Set $X = \{CCC, 1C0, C01\}$ and $i = 1$.
2. $n - |Y| = 15 - (8 + 2 + 2) = 3 \geq 2^{3-2}$, so go to Step 3.
3. The F-set has only a single code word, not two; go to Step 4.
4. There is a set $S(u_1)$, $u_i = 01C$, so $X = \{CCC, 1C0, C01, 01C\}$ and $R = \{0C1, C10, 10C-\}$; go to Step 5.
5. $n - |Y| = 15 - (8 + 2 + 2 + 2) = 1$; go to Step 6.
6. $i = 2$. Since $i > \lfloor 3/2 \rfloor$, go to Step 7.
7. $X = \{CCC, 1C0, C01, 01C, 10C+\}$; stop.

Fifteen code words have been selected, as shown in Table 4.6, for designing the checker part. Figure 4.23(a) shows the PLA implementation of the checker using the code words.

1-out-of-n to k-out-of-2k Code Translator

The translator can be implemented with a PLA by using the following steps:

1. Number the input lines from 1 to n and the output lines from 1 to $2k$ corresponding to a code word X_i from set X.
2. Place a transistor at the intersection of the ith input line and the jth output line if the jth bit position of X_i is at 0.

The translator for 1-out-of-15 code to 3-out-of-6 code is shown in Figure 4.23(b). It should be indicated that the 1-out-of-n checker will be efficient if implemented using PLAs only. Since multilevel random logic is more popular than PLAs in modern VLSI design, there is a need for new design techniques for 1-out-of-n code checkers.

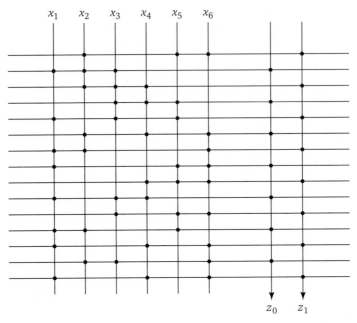

x_1 x_2 x_3 x_4 x_5 x_6

z_0 z_1

(a)

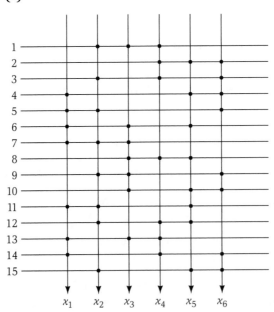

x_1 x_2 x_3 x_4 x_5 x_6

(b)

4.23

Figure

(a) Totally self-checking 3-out-of-6 checker; (b) translator for 1-out-of-5 to 3-out-of-6 code.

4.3   TOTALLY SELF-CHECKING CHECKER FOR BERGER CODE

Over the years several techniques have been presented for designing self-checking checkers for Berger code based on the Type I checker for separable codes proposed by Ashjaee and Reddy [4.12]. Figure 4.24 shows a Type I checker. C_1 is a nonredundant combinational circuit that generates the complements of the check bits from information bits. The two-rail totally self-checking checker circuit compares the k check bits with the outputs of C_1. If there is no fault present in the two-rail checker or in the check bit generator, the outputs of the checker are complementary (i.e., 0 1 or 1 0), otherwise they are 0 0 or 1 1.

Marouf and Friedman [4.13] have presented a procedure for designing totally self-checking checkers for maximal length Berger codes. The combinational circuit C_1 of Figure 4.24 generates the binary number corresponding to the number of 1s in the information bits. It is designed using a set of full-adder modules, which add the information bits $(x_1, x_2, \ldots, x_{2^k-1})$ in the parallel and produce the binary number $g_1, g_2, \ldots, g_k$ corresponding to the number of 1s in the information bits. The number of adder modules required is

$$\sum_{a=1}^{k-1} 2^{a-1}(k-a)$$

where k is the number of check bits.

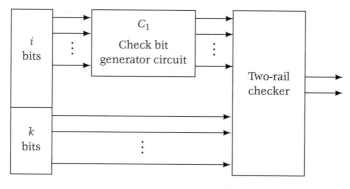

4.24 Figure   Totally self-checking checker for separable codes.

Thus, for Berger codes with information bits $i = 3$ and check bits $k = 2$, the subcircuit C_1 of Figure 4.24 is simply a full-adder module, where the sum output $s = g_1$ and the carry output $c = g_2$. Similarly, for $i = 7, k = 3$, the number of adder modules needed to implement C_1 is 4, and for $i = 15, k = 4$, C_1 needs 11 modules.

The procedure to design subcircuit C_1 consists of the following steps:

1. Let $I = \{x_1, x_2, \ldots, x_i\}$ be the set of i information bits; set $m = k$ and $J = 1$.

2. Let $z = 2^{(m-1)} - 1$.

3. Partition I into three subsets A^J, B^J, and E^J. A^J contains the leftmost z bits, B^J contains the next z bits, and E^J has the rightmost bit.

4. Let $\underline{a}^J (= a^J_{m-1}, a^J_{m-2}, \ldots, a^J_1)$, $\underline{b}^J (= b^J_{m-1}, \ldots, b^J_1)$, and e^J be the binary representation of the number of 1s in the subsets A^J, B^J, and E^J, respectively.

5. Let $\underline{g}^J (= g^J_m, g^J_{m-1}, \ldots, g^J_1)$ be the binary representation of the number of 1s in set I. This is obtained from the following addition:

$$\underline{g}^J = \underline{a}^J + \underline{b}^J + e^J$$

In other words,

$$
\begin{array}{ccccccc}
 & a^J_{m-1} & a^J_{m-2} & \cdots & \cdots & a^J_1 \\
+ & b^J_{m-1} & b^J_{m-2} & \cdots & \cdots & b^J_1 \\
+ & & & & & e^J \\
\hline
g^J_m & g^J_{m-1} & g^J_{m-2} & \cdots & \cdots & g^J_1
\end{array}
$$

where g^J_m is the carry bit. $\underline{g}^J$ is generated using a ripple carry adder with $(m - 1)$ stages.

6. Go to Step 8 if $m = 2$; otherwise, set $m = m - 1$, $L = J$.

7. Let $I = \{A^J\}$, $J = J + 1$; repeat Steps 2–6 to generate $\underline{a}^L = \underline{g}^J \cdot \underline{b}^L$ in an identical manner by making $I = \{B^J\}$.

8. End.

The procedure is illustrated by designing the subcircuit C_1 for the case $i = 7, k = 3$, in steps:

1. $I = \{x_1, x_2, x_3, x_4, x_5, x_6, x_7\}$; $m = 3$, $J = 1$.

2. $z = 2^{(3-1)} - 1 = 3$.

3. $A^1 = \{x_1, x_2, x_3\}$, $B^1 = \{x_4, x_5, x_6\}$, and $E^1 = \{x_7\}$.

4. $\underline{a}^1 = (a^1_2, a^1_1)$, $\underline{b}^1 = (b^1_2, b^1_1)$, $e^1 = x_7$.

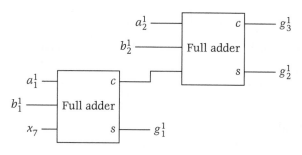

Circuit for g^1.

5. $\underline{g}^1 = (g_3^1, g_2^1, g_1^1)$, where g_3^1, g_2^1, g_1^1 are obtained from

$$
\begin{array}{rcc}
 & a_2^1 & a_1^1 \\
+ & b_2^1 & b_1^1 \\
+ & & x_7 \\
\hline
g_3^1 & g_2^1 & g_1^1
\end{array}
$$

Figure 4.25 shows the generation of $\underline{g}^1$ using a ripple carry adder with $2 (= 3 - 1)$ stages.

6. $m > 2$, $m = 3 - 1 = 2$, and $L = 1$.

7. $I = \{A^1\} = \{x_1, x_2, x_3\}$, $J = 1 + 1 = 2$.

8. $z = 2^{(2-1)} - 1 = 1$ [Step 2].

9. $A^2 = \{x_1\}$, $B^2 = \{x_2\}$, and $E^2 = \{x_3\}$ [Step 3].

10. $\underline{a}^2 = a_1^2 = x_1$, $\underline{b}^2 = b_1^2 = x_2$, $e^2 = x_3$ [Step 4].

11. $\underline{g}^2 = (g_2^2, g_1^2)$, where g_2^2, g_1^2 are obtained from

$$
\begin{array}{rc}
 & x_1 \\
+ & x_2 \\
+ & x_3 \\
\hline
g_2^2 \quad g_1^2 & \text{[Step 5]}
\end{array}
$$

Since $\underline{a}^L = \underline{g}^J$, we have

$$
\underline{a}^1 = \underline{g}^2 \quad \text{or} \quad \left\{ a_2^1, a_1^1 \right\} = \left\{ g_2^2, g_1^2 \right\}
$$

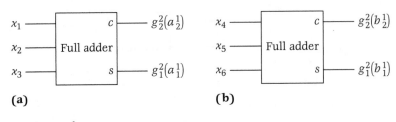

(a) **(b)**

4.26

Figure

Figure 4.26 (a) $\underline{a}^1$; (b) $\underline{b}^1$.

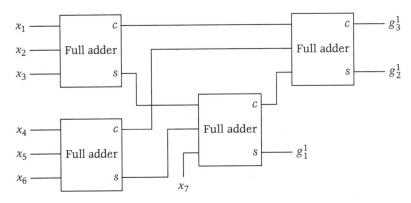

Figure 4.27 The check bit generator circuit for $i = 7, k = 3$ (from [4.13], courtesy of IEEE, ©1978).

In other words, $g_1^2 = a_1^1$ and $g_2^2 = a_2^1 \cdot \underline{a}^1$ can be generated by a full adder as shown in Figure 4.26(a). It can be shown in a similar manner by making $I = \{B^1\} = \{x_4, x_5, x_6\}$ so that

$$\underline{b}^1 = g^2 \quad \text{or} \quad \left\{ b_2^1, b_1^1 \right\} = \left\{ g_2^2, g_1^2 \right\}$$

Figure 4.26(b) shows the generation of $\underline{b}^1$. Since $m = 2$ the procedure is terminated.

Figure 4.27 shows the complete design of the subcircuit C_1 for $i = 7, k = 3$; it is obtained by combining Figures 4.25 and 4.26.

An alternative approach to designing checkers for Berger codes has been proposed by Piestrak [4.14]. In this approach a Berger code is represented as a union of several *m*-out-of-*n* codes. A checker for the Berger code is then implemented using the checkers for the *m*-out-of-*n* codes. This approach requires fewer gates and logic levels than earlier techniques; however, fan-in/fan-out connections within a checker increase with the number of information bits. Moreover, this technique is applicable if the Berger code is derived using a *B*1 encoding scheme and for information bits of length $[2^k - 1, 2^k - 2]$.

Lo and Thanawastien [4.15] proposed a theory of Berger code partitioning. Their partitioning approach is applicable to Berger codes based on *B*1 or *B*0 encoding schemes and information bits of arbitrary length. Rao et al. [4.16] proposed a technique for checker design for Berger codes with 2^k information bits; such checkers have improved testability.

Pierce and Lala [4.17] have proposed a scheme for a check bit generator for Berger codes that is more efficient in terms of gate counts and speed over existing approaches. The check bit generator is designed assuming the *B*1 encoding scheme for non-maximal length Berger code and the *B*0 encoding scheme for maximal length Berger code. The basic building block of the check bit generator is a 4-input 1s-counter (shown in Figure 4.28[1]). The counter has four inputs (a, b, c, d) and generates at its three outputs (out_1, out_2, out_3) the binary representation of the number of 1s in the four input bits.

Two other types of 1s-counters, having two and three inputs, are used to realize checkers for variable length information bits. Figure 4.29 shows the schematics of these counters. The 2-, 3-, and 4-input counters are called the *basic counters*; a 1s-counter of any size can be constructed from them. For example, a 5-input 1s-counter can be constructed from a 3-input 1s-counter, a 2-input 1s-counter, and a standard 2-bit adder.

A check bit generator is implemented with two 1s-counters and an addition array. The information bits are partitioned into two blocks: I_A and I_B. Block I_A contains $\lceil I/2 \rceil$ bits and block I_B contains $I - \lceil I/2 \rceil$ bits. If the number of information bits in a block is not more than four, two appropriate basic counters can be used in conjunction with the appropriate addition array to implement a check bit generator. However, if the number of information bits in a block is more than four, the block is partitioned into two subblocks. The partitioning process is continued until a subblock can be implemented with a predesigned counter (constructed from basic counters) and an addition array.

The number of bits in blocks I_A and I_B is $\log_2(\lceil I/2 \rceil + 1)$ and $\log_2((I - \lceil I/2 \rceil) + 1)$, respectively. The number of 1s in blocks I_A and I_B is represented by vectors A and B, respectively. The check bits k for I information bits is the

---

1. Figures 4.28 through 4.45 are from [4.17], courtesy of Kluwer Academic Publishers, ©1996.

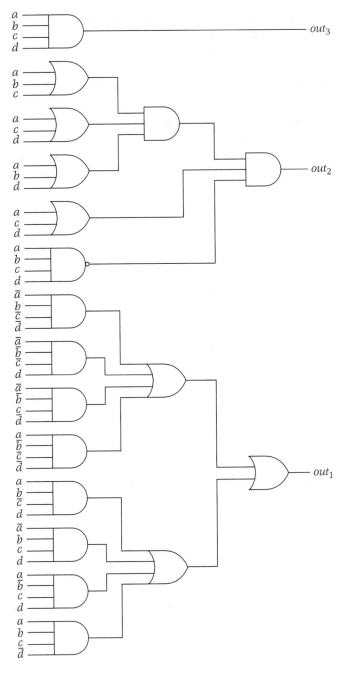

A 4-input 1s-counter.

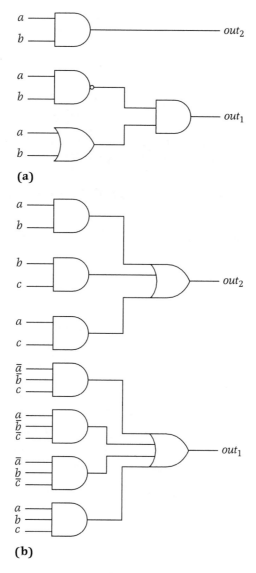

(a) A 2-input 1s-counter; (b) a 3-input 1s-counter.

sum of vectors A and B and is derived by adding two bits at a time as shown below:

$$
\begin{array}{rllllll}
A & = & a_n a_{n-1} & \cdots & \cdots & \cdots & \cdots & a_3 a_2 a_1 a_0 \\
B & = & b_n b_{n-1} & \cdots & \cdots & \cdots & \cdots & b_3 b_2 b_1 b_0 \\
\hline
& & k_n k_{n-1} & \cdots & \cdots & \cdots & \cdots & k_3 k_2 k_1 k_0
\end{array}
$$

If there are 2^n (where n is an integer) inputs to a counter, the counter produces vector A (or B) whose most significant bit is considered to be a *special most significant bit* (SMSB). If the SMSB of A or B is 1, it is the only 1 present in the vector, and a carry is not produced when A and B are added. Let us illustrate it by assuming

number of information bits $I = 7$

then $I_A = 4$ and $I_B = 3$

number of bits in vector $A = 3 \Rightarrow a_2 a_1 a_0$

number of bits in vector $B = 2 \Rightarrow b_1 b_0$

If the information bits in a 4-bit block (e.g., I_A) are all 1s and these bits are the inputs to a 4-bit counter, the most significant bit of the vector (A) (i.e., a_2) will be 1, and a_1 and a_0 will be 0. Therefore, the sum of vector A and another vector B will never produce a carry. The two least significant bits of vectors A and B (i.e., a_0, b_0, a_1, and b_1) are grouped together in a set called a *primary group*. A 2-bit adder is used to sum these bits. The outputs of the adder become the least significant check bits (i.e., k_1 and k_0). The carry-out generated during this addition generally propagates to a secondary group. In other words, the carry becomes the check bit. To illustrate, let us assume

$$I = 1\,0\,1\,1\,0, \quad I_A = 1\,0\,1, \quad \text{and} \quad I_B = 1\,0$$

Therefore,

$$A = a_1 a_0 = 1\,0, \quad B = b_1 b_0 = 0\,1$$

The check bits k are obtained by adding A and B as shown in Figure 4.30. The special 2-bit adder is derived from the standard 2-bit adder shown in Figure 4.31(a). Since $a_1 b_1 = 1\,1$ does not occur in check bit generator design for $((I, k) = 5, 3)$, gates with $a_1 b_1$ in Figure 4.31(a) are discarded to form the special 2-bit adder in Figure 4.31(b). The special 2-bit adder is used when a block or a subblock has five information bits; otherwise the standard 2-bit adder is used to generate the sum of bits in a primary group.

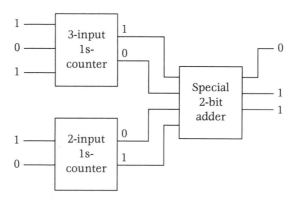

Generation of check bits k.

A *secondary group,* unlike a primary group, may contain a number of bits ranging from 1 to 6. Maximal length information bits ($I = 7, 15, 31, \ldots$) will result in an odd number of bits in the secondary group, whereas non-maximal length information bits will result in an even number of bits. The type of adders to be used to generate the sum of the bits in the secondary group is dependent not only upon the number of bits in the secondary group but also on whether or not the most significant bit of vector A (or B) is an SMSB. When the information bits are non-maximal, the most significant bit of A (or B) may or may not be an SMSB. However, the most significant bit of A is always an SMSB if the information bits are maximal.

We consider several cases with different values of the number of bits in a secondary group. For simplicity we assume the existence of a single secondary group.

Case 1   One bit in the secondary group

	a_2	a_1	a
		b_1	b_0
	k_2	k_1	k_0

This case results only when $I = 7$, therefore a_2 is an SMSB. The carry circuit of the standard 2-bit adder (Figure 4.31(a)) is modified to accommodate a_2 and is shown in Figure 4.32. The sum circuit of the modified 2-bit adder is the same as that of the standard 2-bit adder. The check bit generator block diagram for $(I, k) = (7, 3)$ is shown in Figure 4.33.

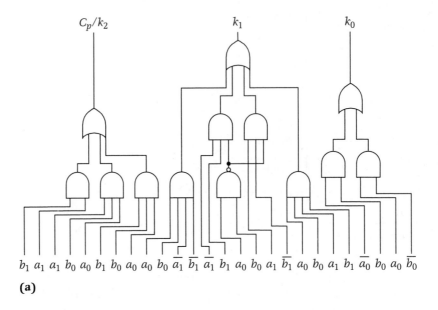

(a)

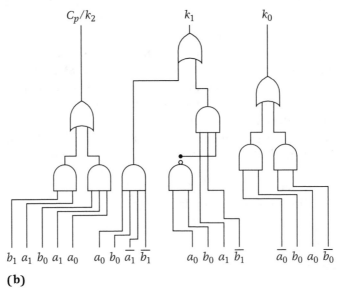

(b)

4.31

Figure

(a) Standard 2-bit adder; (b) special 2-bit adder.

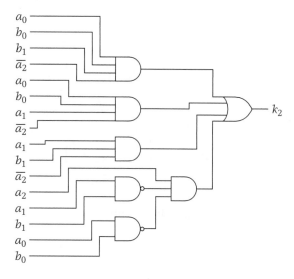

Modified carry circuit of Figure 4.31(a).

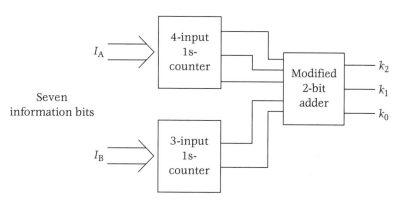

Check bit generator for $(I, k) = (7, 3)$.

To illustrate let us assume

$$I = 1\,1\,1\,1\,1\,0\,1, \ I_A = 1\,1\,1\,1, \ \text{and} \ I_B = 1\,0\,1$$

Therefore,

$$A = 1\,0\,0, \ B = 1\,0, \ k = A + B = 1\,1\,0$$

∎

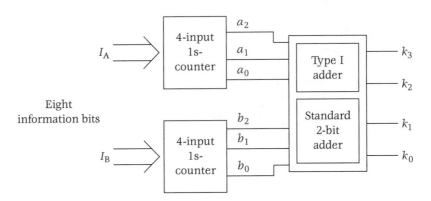

A check bit generator for eight information bits.

Case 2   Two bits in the secondary group

C_p		
a_2	a_1	a_0
b_2	b_1	b_0

| k_3 | k_2 | k_1 | k_0 |

If bits a_2 and b_2 are SMSBs (Figure 4.34), a Type I adder (Figure 4.35) is used to generate check bits k_3 and k_2; otherwise, a full adder is used. Figure 4.36 shows the realization of a full adder.

Let us illustrate Case 2 by designing the check bit generator for the (11,4) Berger code.

$$I_A = 6 = (\lceil I/2 \rceil) \text{ bits}$$

$$I_B = 5 = (I - \lceil I/2 \rceil) \text{ bits}$$

Therefore,

$$A = a_2a_1a_0, \ B = b_2b_1b_0, \ \text{and } k = k_3k_2k_1k_0$$

The implementation of the check bit generator for the (11,4) code is shown in Figure 4.37. Note that neither a_2 nor b_2 is an SMSB; therefore, a full adder rather than the Type I adder is used to sum the bits in the secondary group.

∎

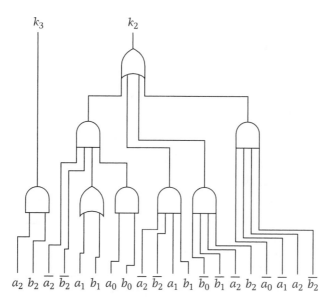

4.35
Figure

Type 1 adder.

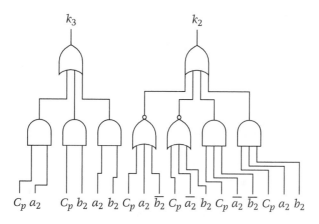

4.36
Figure

Full adder.

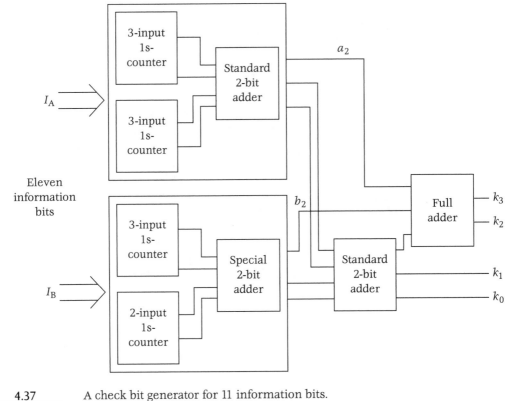

4.37

Figure A check bit generator for 11 information bits.

Case 3 Three bits in the secondary group

	C_p		
a_3	a_2	a_1	a_0
	b_2	b_1	b_0
k_3	k_2	k_1	k_0

This case results only when $I = 15$, therefore a_3 is an SMSB (see Figure 4.38). The carry circuit of the full adder is modified to generate check bit k_3. The modification of the full adder's carry circuit is shown in Figure 4.39. ∎

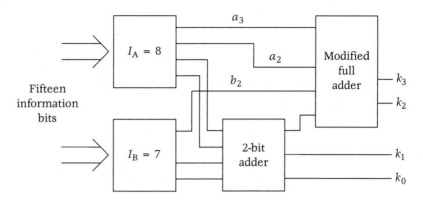

4.38

Figure

A check bit generator for 15 information bits.

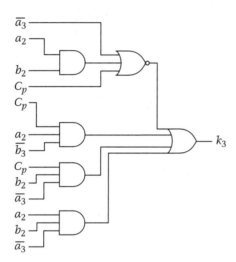

4.39

Figure

Modified carry circuit of Figure 4.36.

Case 4   Four bits in the secondary group

	C_p		
a_3	a_2	a_1	a_0
b_3	b_2	b_1	b_0
k_3	k_2	k_1	k_0

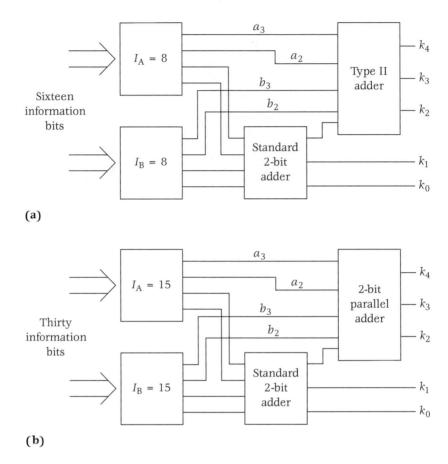

(a)

(b)

(a) A check bit generator for 16 information bits; (b) a check bit generator for 30 information bits.

If both bits a_3 and b_3 are SMSBs, then a Type II adder (Figure 4.40(a)) is used; otherwise a 2-bit parallel adder (Figure 4.40(b)) is used. The circuit realization of the Type II adder is shown in Figure 4.41. The structure of the 2-bit parallel adder shown in Figure 4.42 consists of two standard 2-bit adders, one of them with a carry-in of 1, and three 2-to-1 multiplexers. The select lines of the multiplexers are driven by the carry-out of the primary group. For example, the outputs of the 2-bit adder (with $C_p = 1$) is chosen when the primary group produces a carry-out of 1. The realization of the 2-bit adder (with $C_p = 1$) is shown in Figure 4.43. ■

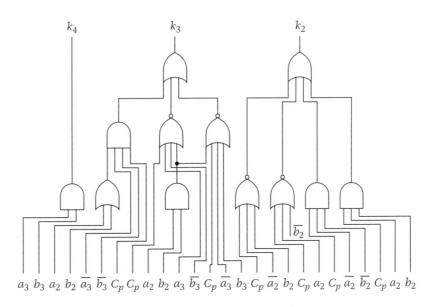

$$a_3\ b_3\ a_2\ b_2\ \overline{a_3}\ \overline{b_3}\ C_p\ C_p\ a_2\ b_2\ a_3\ \overline{b_3}\ C_p\ \overline{a_3}\ b_3\ C_p\ \overline{a_2}\ b_2\ C_p\ a_2\ C_p\ \overline{a_2}\ \overline{b_2}\ C_p\ a_2\ b_2$$

4.41	A Type II adder.
Figure	

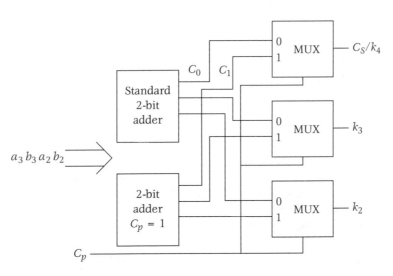

4.42	A 2-bit parallel adder.
Figure	

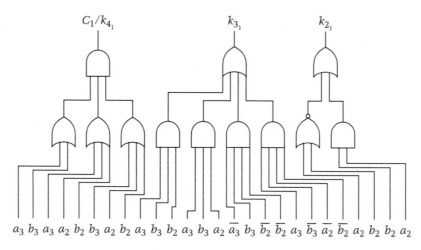

$a_3\ b_3\ a_3\ a_2\ b_2\ b_3\ a_2\ b_2\ a_3\ b_3\ b_2\ a_3\ b_3\ a_2\ \overline{a_3}\ b_3\ \overline{b_2}\ \overline{b_2}\ a_3\ \overline{b_3}\ \overline{a_2}\ \overline{b_2}\ a_2\ b_2\ b_2\ a_2$

4.43

Figure

A 2-bit adder (with $C_p = 1$).

If a secondary group has four bits, it is treated as a primary group. The selection of the adder modules needed to add bits on subsequent secondary groups is chosen using the guidelines discussed in Cases 1–4. However, two special cases may arise when more than one secondary group is considered.

Case 5   One bit in the secondary group 2

C_S					
a_4	a_3	a_2	|	a_1	a_0
	b_3	b_2	|	b_1	b_0

| k_4 | k_3 | k_2 | | k_1 | k_0 |

In this case the configuration of the 2-bit parallel adder (Figure 4.42) is modified to accommodate the SMSB a_4. This is done by replacing the standard 2-bit parallel adder by the modified 2-bit adder in Case 1; bit a_4 becomes an input to the modified 2-bit adder.   ■

Case 6   Two bits (both SMSBs) in the secondary group 2

C_S					
a_4	a_3	a_2	|	a_1	a_0
b_4	b_3	b_2	|	b_1	b_0

| k_5 | k_4 | k_3 | k_2 | | k_1 | k_0 |

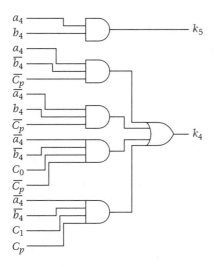

4.44

Figure

A Type III adder.

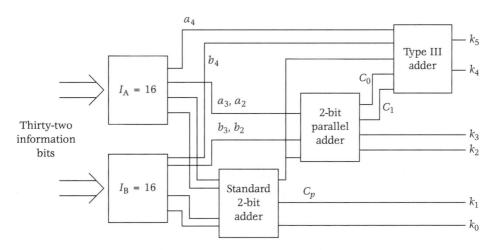

4.45

Figure

A check bit generator for 32 information bits.

Since a_4 and b_4 are SMSBs, a Type III adder (shown in Figure 4.44) is used to generate check bits k_5 and k_4. Note that in this case carries C_0 and C_1 generated by the 2-bit parallel adder propagate to the subsequent secondary group. Otherwise, as in Case 2, a full adder is used to generate the sum of bits C_5, a_4, and b_4. The check bit generator for (32, 6) Berger code is shown in Figure 4.45. ∎

4.4   TOTALLY SELF-CHECKING CHECKER FOR LOW-COST RESIDUE CODE

The check bits b for a low-cost residue code are obtained by performing a modulo-$(2^b - 1)$ addition of the information bits [4.17]. This is usually obtained by dividing the number of information bits into k groups, each containing b bits. A modulo-$(2^b - 1)$ addition of the k b-bit groups results in the check bits. For example, let us calculate the check bits for information bits $1\,0\,0\,1\,1\,1\,1\,0$, assuming $b = 2$. The information bits must then be divided into four 2-bit groups. The groups are added together in modulo 3 ($= 2^2 - 1$) binary adders with end-around carry. This technique is illustrated in Figure 4.46(a). Using this technique the check bits for the given information bits are derived as follows:

$$10 + 01 + 11 + 10$$
$$= 11 + 10$$
$$= 10$$

If the number of information bits is not divisible by b, one of the adders gets less than $2b$ inputs; the missing bits are assumed to be 0s. For example, the check bits for information bits $1\,0\,1\,0\,1\,1\,0\,0\,1\,0\,1$ are generated by setting the missing bits equal to 0 as shown in Figure 4.46(b).

Ashjaee and Reddy [4.12] have proposed a totally self-checking checker for low-cost residue codes. As suggested by Figure 4.47, the checker compares the check bits of a code word with new check bits generated from the information bits of the code word. A disagreement between the two values is detected as a fault. The check bit generator is in fact a tree network of modulo-$(2^b - 1)$ adders. Under fault-free conditions, the check bits C_S^* are equal to C_S. The totally self-checking two-rail comparator checks whether C_S and $\overline{C_S^*}$ are bit-by-bit complements of each other or not.

However, in certain cases C_S^* may not be equal to C_S even under fault-free operation. For example, in the adder circuit of Figure 4.47 let $I_A = 1\,1\,1\,0$ and $I_B = 0\,0\,1\,0$, then $C_A = 1\,0$, $C_B = 1\,0$ assuming $b = 2$. If 2s-complement addition is performed, $I_S = 0\,0\,0\,0$, $C_S = 0\,1$, and $C_S^* = 0\,0 \neq C_S$. The reason for this is that $C_S = C_S^*$ (for all I_A and I_B) if 1s-complement addition is assumed and the number of information bits is divisible by b. If 2s-complement or sign-magnitude arithmetic is used, the carry-out bit of the binary adder must be included as the most significant position of the sum bits (as shown by the dotted line in Figure 4.47). In this case $I_S = I_A + I_B = 1\,0\,0\,0\,0$ instead of $0\,0\,0\,0$ and $C_S^* = 0\,1 = C_S$ when there is no fault.

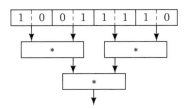

(a)

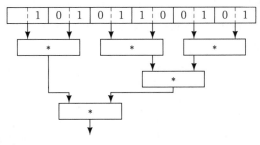

(b)

4.46
Figure

Check bit generators for low-cost residue codes. * indicates mod-3 adders.

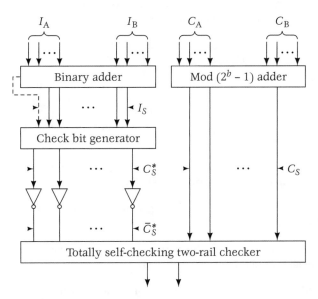

4.47
Figure

A totally self-checking checker for low-cost residue codes (adapted from [4.12]).

4.5   REFERENCES

[4.1]   Rao, T. R. N., and E. Fujiwara. *Error-Control Coding for Computer Systems.* Prentice Hall, 1989.

[4.2]   Anderson, D. A., and G. Metze. "Design of totally self checking check circuits for *m*-out-of-*n* codes." *IEEE Trans. Computers,* C-22 (March 1973): 263–269.

[4.3]   Lo, J. C. "A novel area-time efficient static CMOS totally self-checking comparator." *IEEE Jour. of Solid-State Circuits* 28, no. 2 (February 1993): 165–168.

[4.4]   Reddy, S. M. "A note on self-checking checkers." *IEEE Trans. Computers,* C-23 (October 1974): 1100–1102.

[4.5]   Reddy, S. M., and J. R. Wilson. "Easily testable cellular realizations for the (exactly *p*)-out-of-*n* and (*p* or more)-out-of-*n* logic functions." *IEEE Trans. Computers,* C-23 (January 1974): 98–100.

[4.6]   Smith, J. E. "The design of totally self checking check circuits for a class of unordered codes." *Jour. Design Automation and Fault-Tolerant Computing* 2 (October 1977): 321–343.

[4.7]   Marouf, M. A., and A. D. Friedman. "Efficient design of self-checking checker for any *m*-out-of-*n* code." *IEEE Trans. Computers,* C-27 (June 1978): 482–490.

[4.8]   Lala, P. K., F. Busaba, and M. Zhao. "Transistor-level implementation of totally self-checking checkers for a subset of *m*-out-of-2*m* codes." *Proc. 2nd IEEE International On-Line Testing Workshop*, France, July 1996.

[4.9]   Tao, D. L., and P. K. Lala. "Three level totally self-checking checker for 1-out-of-*n* code." *Proc. 17th Fault Tolerant Computing Conf.*, 1987, 108–113.

[4.10]   Tao, D. L. "Application of coding techniques in the design of self-checking PLAs." Ph.D thesis, Syracuse University, 1988.

[4.11]   Wang, S. L., and A. Avizienis. "The design of totally self-checking circuits using programmable logic arrays." *Proc. Int. Symp. Fault-Tolerant Computing*, 1978, 179–184.

[4.12]   Ashjaee, M. J., and S. M. Reddy. "On totally self-checking checkers for separable codes." *IEEE Trans. Computers,* C-26 (August 1977): 737–744.

[4.13]   Marouf, M. A., and A. D. Friedman. "Design of self-checking checkers for Berger codes." *Proc. Int. Symp. Fault-Tolerant Computing*, 1978, 179–184.

[4.14]   Piestrak, S. J. "Design of fast self-checking checkers for a class of Berger codes." *IEEE Trans. Computers,* C-36 (May 1987): 629–634.

[4.15]   Lo, J. C., and S. Thanawastien. "The design of fast totally self-checking Berger checkers based on Berger code partitioning." *Proc. 18th Int. Symp. Fault-Tolerant Computing,* June 1988, 226–231.

[4.16] Rao, T. R. N., G. L. Feng, M. S. Kolluru, and J. C. Lo. "Novel totally self-checking Berger code checker design based on generalized Berger code partitioning." *IEEE Trans. Computers* 2 (August 1993): 1020–1024.

[4.17] Pierce, D. A., and P. K. Lala. "Modular implementation of efficient self-checking checkers for the Berger code." *Jour. Electronic Testing: Theory and Applications* 9 (1996): 279–294.

[4.18] Wakerly, J. *Error Detecting Codes, Self Checking Circuits and Applications.* Elsevier North-Holland, 1978.

5

Self-Checking Sequential Circuit Design

A sequential circuit, unlike a combinational circuit, has memory. In other words, the outputs of a sequential circuit depend not only on the present values of the inputs but also on the past inputs. The behavior of a sequential circuit is described by an abstract model known as a *finite state machine* or simply *state machine*. If the next state of a state machine is determined uniquely by the present state and the present input, then the machine is considered to be *deterministic* [5.1]. In general, the output of a state machine is a function of the present state and the inputs. A state machine possessing this feature is known as a *Mealy machine*. Alternatively, if the output of a state machine is a function of only the present states and is not dependent upon external inputs, it is known as the *Moore machine*. A Mealy machine is more general than a Moore machine. Figure 5.1 shows the block diagram of a Mealy machine. It consists of a combinational logic block and memory elements to hold the state information. A *state* is an n-bit vector where n is the number of memory elements (latches and flip-flops) in the state machine. Each state has a unique vector representing it. A state is considered to be valid if it can be reached from the reset state of the machine by using an input sequence of arbitrary length; otherwise the state is invalid [5.2].

The behavior of a state machine can be represented by a *state transition graph* (STG). An STG has as many vertices as there are states in the state machine. A set of edges is used to join the vertices, each edge corresponding to a transition from one state to another. The labels on the edges identify the input causing the transition and the output produced during the transition. Figure 5.2 shows an STG for a three-state machine. A state machine is said to

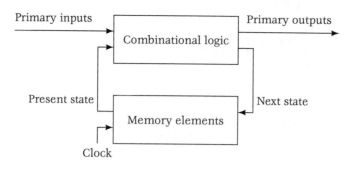

A Mealy-type state machine.

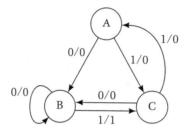

A state transition graph.

be *fully specified* if the next state and the output for each state of the machine are specified for all possible input combinations. Otherwise, the machine is *incompletely specified*.

5.1   FAULTS IN STATE MACHINES

Faults in state machines can occur externally, that is, at the primary inputs or outputs of a machine, or inside the combinational logic block corresponding to the output and next state logic. A fault in a state machine is considered to be redundant if the output responses of the faulty and the fault-free machine are identical for all possible input sequences. Redundant faults in sequential logic circuits can be categorized as follows [5.2]:

+ Combinationally redundant faults (CRF)
+ Sequentially redundant faults (SRF)

A fault is *combinationally redundant* if its effect cannot be observed at primary outputs or if it does not affect next states for any input pattern and any present state. *Sequentially redundant* faults can be classified into three categories [5.3].

1. *Equivalent SRF:* Such a fault makes a circuit move to a next state or creates a next state that is equivalent to the expected next state.

   To illustrate, let us consider the state diagram shown in Figure 5.3(a). Assume a fault has modified the state diagram as shown in Figure 5.3(b), where the edge from state A, instead of going to state C, goes to state D.

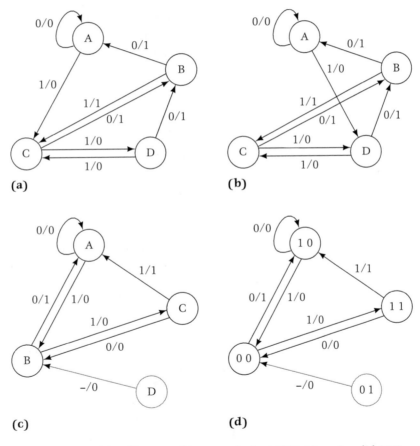

5.3

Figure

(a) A state transition diagram; (b) an equivalent SRF; (c) an invalid SRF; (d) an isomorph SRF.

Since states C and D are equivalent, the output response produced by the fault-free and the faulty circuits is identical. Thus, the assumed fault in this case is an equivalent SRF.

2. *Invalid SRF:* This is a fault that does not cause erroneous transitions from a valid state to any other valid state in a state diagram, but may cause a transition from an invalid state to a valid state. Since the circuit will never be in an invalid state, such a fault will have no effect on the circuit operation and hence is redundant.

   For example, in Figure 5.3(c) state D is nonvalid, and a corrupted edge is created from D to B due to a fault. Since the circuit will never be in state D, this fault is redundant.

3. *Isomorph SRF:* This is a fault that changes one or more state encodings in a sequential circuit, but the faulty circuit is isomorphic to the original circuit. Such a fault is illustrated with the state diagram shown in Fig. 5.3(c), in which the codes for states A, B, C, and D are assumed to be 0 0, 1 0, 0 1, and 1 1, respectively. If the codes for states A and B and states C and D are interchanged, the resulting STG shown in Figure 5.3(d) is isomorphic to Figure 5.3(c).

5.2

SELF-CHECKING STATE MACHINE DESIGN TECHNIQUES

The concepts of self-testing and fault-secureness given in Chapter 3 can be extended to state machines [5.4]. A state machine is self-testing if, for every fault in a fault set, there is a code space input/state pair in the circuit such that a non-code space output is produced. A state machine is fault-secure if, for every fault from a faulty set, the machine never produces an incorrect code space output for code space inputs. Thus, it is possible for the machine to pass through several incorrect states maintaining correct code space output before an error indication is given. A state machine is totally self-checking if it is both self-testing and fault-secure.

The problem of designing totally self-checking state machines has been examined by Diaz and Azema [5.5] and Ozguner [5.6].

Diaz and Azema described a procedure for designing Moore-type state machines. The first step of the procedure is to select code words S_x, S_y, and S_z, which are used during normal operation of the machine by the combinations

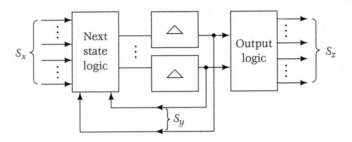

5.4

Figure

A self-checking sequential machine model (from [5.5], courtesy of IEEE, ©1979).

of primary inputs, state variables, and outputs, respectively. If the machine has m inputs, p outputs, and n states, then

$\{S_x\} \subseteq \{m\text{-out-of-}2m \text{ code words}\}$

$\{S_y\} \subseteq \{n/2\text{-out-of-}n \text{ code words}\}$

and

$\{S_z\} \subseteq \{p\text{-out-of-}2p \text{ code words}\}$

Figure 5.4 shows a model of the proposed self-checking sequential machine.

If the states of a machine are coded by using a k-out-of-n code, it is possible to express Y_i, the next state corresponding to y_i, by monotonic equations (which require only uncomplemented variables)

$$Y_i = \sum_j I_j(M_{j1} + \cdots + M_{jp})$$

with

$$M_{jn} = y_{n1} \cdots y_{nk}$$

where $y_{n1} \ldots y_{nk}$ are the variables having the value 1 in a present state q_n that is transferred by input I_j to a state with $Y_i = 1$. This is called an *on-set realization*.

Each output Z is represented by a pair $(Z_1, \overline{Z}_1) \in S = \{(0, 1), (1, 0)\}$, when the machine is fault-free. Then, as the machine is of the Moore type, the outputs

	Input		Output
Present state	$x = 0$	$x = 1$	Z
A	C	A	1
B	D	C	1
C	B	D	0
D	C	A	0

Next state

(a)

	I_1	I_2	
$y_4 y_3 y_2 y_1$	0 1	1 0	$Z \overline{Z}$
0011	1010	0011	10
0101	1001	1010	10
1010	0101	1001	01
1001	1010	0011	01

(b)

5.5

Figure

A Moore-type state machine: (a) a state table; (b) input, state, and output assignments.

can be expressed as functions of the states. Hence for each state q coded by $y_{n1} \ldots y_{nk}$,

$$Z = f(q) = 1 \Rightarrow \begin{cases} Z_1 = y_{n1} \ldots y_{nk} \\ \overline{Z}_1 = 0 \end{cases}$$

$$Z = f(q) = 0 \Rightarrow \begin{cases} Z_1 = 0 \\ \overline{Z}_1 = y_{n1} \ldots y_{nk} \end{cases}$$

This realization can easily be extended to the set of states and the set of outputs in a machine and is called *on-set realization of outputs*. As an example, consider the state table of the Moore-type machine shown in Figure 5.5(a); the encoded state table is shown in Figure 5.5(b).

From Figure 5.5(b), it can be derived that

$$S_x = \{0\,1, 1\,0\}, \quad S_y = \{0\,0\,1\,1, 0\,1\,0\,1, 1\,0\,1\,0, 1\,0\,0\,1\}, \quad \text{and} \quad S_z = \{0\,1, 1\,0\}$$

The next state and output equations can be written in on-set form as follows:

$$Y_1 = I_1 y_1 y_3 + I_1 y_2 y_4 + I_2 y_1 y_2 + I_1 y_2 y_4 + I_2 y_1 y_4$$

$$Y_2 = I_1 y_1 y_2 + I_1 y_1 y_4 + I_2 y_1 y_2 + I_2 y_1 y_3 + I_2 y_1 y_4$$

$$Y_3 = I_1 y_2 y_4$$

$$Y_4 = I_1 y_1 y_2 + I_1 y_1 y_3 + I_1 y_1 y_4 + I_2 y_1 y_3 + I_2 y_2 y_4$$

$$Z = y_1 y_2 + y_1 y_3$$

$$\overline{Z} = y_2 y_4 + y_1 y_4$$

The next state and output equations are realized using two-level AND-OR logic. Each implicant in the above expressions is a test for the corresponding AND gate and the output of the OR gate, for stuck-at-0 faults. The significance of the arrows is explained later.

Let M be the set of implicants, corresponding to the input vectors $(S_x \times S_y)$, applied to the next state logic, and M_i be the set of implicants for $Y_i (i = 1..n)$. Then

$$M = \{I_1 y_1 y_2,\ I_1 y_1 y_3,\ I_1 y_1 y_4,\ I_2 y_1 y_3,\ I_2 y_2 y_4,\ I_1 y_1 y_2,\ I_2 y_1 y_4,\ I_1 y_2 y_4\}$$

$$M_1 = \{I_1 y_1 y_3,\ I_1 y_2 y_4,\ I_2 y_1 y_2,\ I_2 y_2 y_4,\ I_2 y_1 y_4\}$$

$$M_2 = \{I_1 y_1 y_2,\ I_1 y_1 y_4,\ I_2 y_1 y_2,\ I_2 y_1 y_3,\ I_2 y_1 y_4\}$$

$$M_3 = \{I_1 y_2 y_4\}$$

$$M_4 = \{I_1 y_1 y_2,\ I_1 y_1 y_3,\ I_1 y_1 y_4,\ I_2 y_1 y_3,\ I_2 y_2 y_4\}$$

Hence

$$M - M_1 = \{I_1 y_1 y_2,\ I_1 y_1 y_4,\ I_2 y_1 y_3\}$$

$$M - M_2 = \{I_1 y_1 y_3,\ I_2 y_2 y_4,\ I_1 y_2 y_4\}$$

$$M - M_3 = \{I_1 y_1 y_2,\ I_1 y_1 y_3,\ I_1 y_1 y_4,\ I_2 y_1 y_3,\ I_2 y_2 y_4,\ I_2 y_1 y_2,\ I_2 y_1 y_4\}$$

$$M - M_4 = \{I_2 y_2 y_4,\ I_2 y_1 y_2,\ I_2 y_1 y_4,\ I_1 y_2 y_4\}$$

A "divider" of an implicant, with respect to one of the variables appearing in it, is obtained by making the variable equal to 1 in the implicant. For instance, the divider of the implicant $I_1 y_2 y_4$ with respect to I_1 is $y_2 y_4$. This divider means that in order to test I_1 stuck-at-1, an input combination in which $y_2 = y_4 = 1$ is required.

In order to find the untested AND gate inputs, each implicant in $M_i (i = 1, n)$ is considered successively. If the divider of an implicant with respect to one of its variables is not included in one implicant of $\{M - M_i\}$, the input of the AND gate corresponding to that variable is not tested. For example, the divider of implicant $I_1 y_2 y_4$ with respect to I_1 (i.e., $y_2 y_4$) is not included in any implicant of $\{M - M_1\}$; therefore, input I_1 of the AND gate realizing $I_1 y_2 y_4$ is not tested. The untested inputs in Y_1, Y_2, Y_3, and Y_4 can be found in this manner and are marked with arrows.

The self-testing functions $Y_{1(ST)}$, $Y_{2(ST)}$, $Y_{3(ST)}$, and $Y_{4(ST)}$ are obtained by removing the untested inputs:

$$Y_{1(ST)} = I_1 y_3 + y_2 y_4 + I_2 y_2 + I_2 y_4$$

$$Y_{2(ST)} = y_1 y_2 + y_1 y_4 + I_2$$

$$Y_{3(ST)} = I_1 y_2 y_4$$

$$Y_{4(ST)} = I_1 y_1 + I_2 y_2 y_4 + y_3$$

The untested inputs in the output logic can be found in the same manner:

$$M = \{y_1 y_2, \ y_2 y_4, \ y_1 y_3, \ y_1 y_4\}$$

$$M - M_1 = \{y_2 y_4, \ y_1 y_4\}$$

$$M - M_2 = \{y_1 y_2, \ y_1 y_3\}$$

Hence

$$Z = y_1 y_2 + y_1 y_3 \quad \overline{Z} = y_2 y_4 + y_1 y_4$$

The output logic is not self-checking, so an extra self-checking checker for 2-out-of-4 has to be added to the output logic. The totally self-checking realization of the output logic is shown in Figure 5.6.

Ozguner presented a method for designing totally self-checking Mealy-type state machines. In principle her method is very similar to that of Diaz and Azema; it uses 1-out-of-n code for state assignment and 1-out-of-p code for the input states. The outputs can be encoded with any code in which the code words have no ordering relation. As an example, let us consider the state machine shown in Figure 5.7(a). It has $n = 4$ internal states and $p = 4$ input states. Hence the internal states and the input states can be encoded using 1-out-of-4 code; the outputs are encoded using 1-out-of-2 code. The encoded state table is shown in Figure 5.7(b).

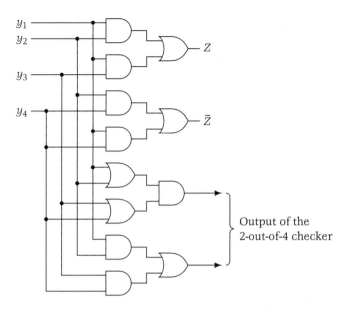

5.6

Figure

Totally self-checking realization of the output logic of Figure 5.5.

Input

State	$\bar{x}_1\bar{x}_2$	$\bar{x}_1 x_2$	$x_1\bar{x}_2$	$x_1 x_2$
S_1	$S_2, 0$	$S_2, 1$	$S_4, 1$	$S_1, 1$
S_2	$S_2, 0$	$S_3, 0$	$S_4, 1$	$S_1, 1$
S_3	$S_1, 0$	$S_4, 0$	$S_3, 0$	$S_1, 1$
S_4	$S_2, 1$	$S_1, 0$	$S_4, 0$	$S_1, 0$

(a)

	I_1	I_2	I_3	I_4
$y_4 y_3 y_2 y_1$	0 0 0 1	0 0 1 0	0 1 0 0	1 0 0 0
0 0 0 1	0 0 1 0, 0 1	0 0 1 0, 1 0	1 0 0 0, 1 0	0 0 0 1, 1 0
0 0 1 0	0 0 1 0, 0 1	0 1 0 0, 0 1	1 0 0 0, 1 0	0 0 0 1, 1 0
0 1 0 0	0 0 0 1, 0 1	1 0 0 0, 0 1	0 1 0 0, 0 1	0 0 0 1, 1 0
1 0 0 0	0 0 1 0, 1 0	0 0 0 1, 0 1	1 0 0 0, 0 1	0 0 0 1, 0 1

(b)

5.7

Figure

(a) A state table; (b) code assignments.

The next state and output equations are written in the on-set form as follows:

$$Y_1 = I_1 y_3 + I_2 y_4 + I_4 y_1 + I_4 y_2 + I_4 y_3 + I_4 y_4$$

$$Y_2 = I_1 y_1 + I_1 y_2 + I_1 y_4 + I_2 y_1$$

$$Y_3 = I_2 y_2 + I_3 y_3$$

$$Y_4 = I_2 y_3 + I_3 y_1 + I_3 y_2 + I_3 y_4$$

$$Z = I_1 y_4 + I_2 y_1 + I_3 y_1 + I_3 y_2 + I_4 y_1 + I_4 y_2 + I_4 y_3$$

$$\overline{Z} = I_1 y_1 + I_1 y_2 + I_1 y_3 + I_4 y_4 + I_2 y_2 + I_2 y_3$$
$$+ I_2 y_4 + I_3 y_3 + I_3 y_4$$

The output circuit uses the same AND gates as the excitation circuit. Any stuck-at-0 fault in the excitation circuit causes the next state to be all 0s; the output will produce a non-code output at the next clock pulse. A stuck-at-0 fault in the OR gates of the output circuit also causes a non-code output. Any stuck-at-0 fault on the input connections causes the outputs to be all 0s. All single stuck-at-1 faults in the excitation circuit, except those on AND inputs, are equivalent to single stuck-at-1 faults on state variables. If the state variable y_k is stuck-at-1, then for input I_j and any state S_i the output of the machine will be

$$Z(I_j, S_k) \vee Z(I_j, S_i)$$

This is because y_k at logic 1 corresponds to state S_k. For example, if y_4 is stuck-at-1 and the machine is in state S_1 (or S_2, S_3), the output in response to I_1 will be a non-code output:

$$Z(I_1, S_1) \vee Z(I_1, S_4)$$
$$= 0\,1 \vee 1\,0 = 1\,1$$

Similarly, if the input I_j is stuck-at-1, then for state S_k and any input I_i the output will be

$$Z(S_k, I_j) \vee Z(S_k, I_i)$$

For example, if input I_1 is stuck-at-1 and the machine is in state S_2, the output in response to I_3 (or I_4) is

$$Z(S_2, I_1) \vee Z(S_2, I_2)$$
$$= 0\,1 \vee 1\,0 = 1\,1$$

In general state machines designed by this method will produce non-code output in the presence of faults provided the following two conditions are met:

1. $\forall (S_k, I_j)\ k = 1 \ldots n,\ j = 1 \ldots p\ \exists (S_i, I_j), i \neq k$
such that $Z(S_i, I_j) \neq Z(S_k, I_j)$

2. $\forall (S_k, I_j)\ k = 1 \ldots n,\ j = 1 \ldots p\ \exists (S_k, I_i), i \neq j$
such that $Z(S_k, I_i) \neq Z(S_k, I_j)$

It would be necessary to add additional outputs to machines that do not satisfy these rules. Since the output circuit is realized using the same AND gates as the excitation circuit, the additional logic will be an OR gate for each extra output.

Both of the designs presented are totally self-checking for single and uni-directional multiple faults. The advantage of Ozguner's method is that it does not require the addition of a totally self-checking checker, if the sequential machine does not have a code-disjoint output.

Jha and Wang [5.7] have also proposed a scheme for designing self-checking Mealy-type state machines. In this scheme the outputs of a state machine are encoded using the Berger code, and the states are encoded using the m-out-of-n code. The m-out-of-n code is selected such that

$$\binom{n}{m} \geq s \quad \text{and} \quad m \leq \lfloor n/2 \rfloor$$

where s is the number of states in the machine. The output and the next state logic circuits are designed such that there are inverters only at the circuit inputs, not internal to the circuit.

Let us illustrate the application of the scheme by applying it to the state table shown in Figure 5.8(a). Since there are five states, we can satisfy the above conditions by choosing 2-out-of-4 code to encode the states, as shown in Figure 5.8(b). The outputs are encoded as follows. First the number of 0s in the output patterns are computed. Then output patterns are grouped such that all patterns with the same number of 0s are included in one group. If there are G such groups and r is the number of check bits, then

$$r = \lceil \log_2(G + 1) \rceil$$

The r check bits are assigned to different output groups such that a group with the largest number of 0s receives the highest value of r; the group with the next largest zero-count receives the next highest value of r. An all-1s check bit pattern is not used. For example, a set of output patterns and their zero-counts are shown in Figure 5.9(a). Based on the zero-counts, there are four

Present state	Input x = 0	x = 1
A	E,0 0	B,1 0
B	D,0 1	A,0 0
C	C,0 0	D,0 1
D	B,0 1	E,1 0
E	A,1 0	C,0 0

(a)

	y_1	y_2	y_3	y_4
A	1	1	0	0
B	0	1	1	0
C	0	0	1	1
D	1	0	0	1
E	1	0	1	0

(b)

5.8 Figure (a) A state table; (b) state encoding using 2-out-of-4 code.

Output bits	Zero count	Group
0 0 0 0 0 0	5	G_1
0 0 0 1 0 0	4	G_2
1 0 0 0 0 0		
0 1 0 0 0 0		
0 0 1 1 1 1	2	G_3
1 1 0 0 1 1		
1 0 1 1 1 1	1	G_4
0 1 1 1 1 1		
1 1 1 1 1 0		

(a)

Outputs	Check bits
0 0	1 0
0 1	0 1
1 0	0 1

(b)

5.9 Figure (a) Grouping of outputs; (b) output encoding.

output groups as identified in Figure 5.9(a). Therefore, $G = 4$ and hence $r = 3$. A possible assignment of check bits to output patterns is to append 1 0 1, 1 0 0, 0 1 1, and 0 1 0 to groups G_1, G_2, G_3, and G_4, respectively.

Based on the above discussion the outputs for the state machine of Figure 5.8 can be encoded as shown in Figure 5.9(b).

5.3   ELIMINATION OF BIDIRECTIONAL ERRORS

The detection and elimination of faults that might cause bidirectional errors was discussed in Chapter 3. It was shown that if a prime and irredundant combinational circuit is designed such that all faults at the inputs that created bidirectional error at the output are removed, any fault in the circuit, internal or at the inputs, can only result in either single-bit or unidirectional multibit errors at the output irrespective of the way the circuit is implemented. Also, it was shown that if no input cubes are x-bidirectional, then any fault at an input line can only create unidirectional errors at the outputs.

Extending the above results, if the inputs to a circuit are m-out-of-n code, then there are no x-bidirectional input cubes, since the minimum distance between any two code words is at least 2. Therefore, a single stuck-at fault at an input to the circuit can create only a unidirectional error. Also, any unidirectional multiple faults at the inputs can cause either a single-bit error or unidirectional multibit errors at the outputs. This can be explained as follows. Let us assume that C is a fault-free code word at the input. A unidirectional multiple fault at the input will result in a non-code word C_f. C will cover C_f if the unidirectional fault consists of multiple stuck-at-0 faults, and C is covered by C_f in case the unidirectional fault consists of multiple stuck-at-1 faults. For example, if the input code word $C = 0\,0\,1\,1$, and a unidirectional multiple fault at the input changes C to $C_f = 0\,0\,0\,0$, then C covers C_f. On the other hand, if the unidirectional multiple fault changes C to $C_f = 1\,1\,1\,1$, then C is covered by C_f.

Let us consider both possibilities:

Case 1 C covers C_f. Hence, C_f cannot cover any of the input code words. The output corresponding to an input code word covering C_f will either be the expected correct code word or will be all 0s.   ■

Case 2 C_f covers C. In this case, either the output for an input code word covered by C_f will be the expected correct output, or there will be one or more 0 to 1 error(s) in the output bits.   ■

In other words, if the inputs to a circuit are encoded using m-out-of-n code, a unidirectional multiple fault at the inputs can only cause either a single-bit error or a unidirectional multiple error at the outputs. To illustrate let us consider the following truth table, which has 3-out-of-6 code words as inputs.

Input	Output
$a\ b\ c\ d\ e\ f$	$w\ x\ y$
1 0 0 1 0 1	0 0 1
0 1 1 0 1 0	1 1 0
1 1 0 0 0 1	0 1 0
0 0 0 1 1 1	1 0 1

The Boolean expressions for the output bits are

$$w = bce + def$$
$$x = bce + abf$$
$$y = adf + def$$

Let us consider Case 1 first, and assume input bits a and b change to 0s, that is, there is a 2-bit unidirectional error at the primary inputs. This will result in each faulty code word being covered by the corresponding fault-free code word. The outputs for the faulty code words are shown below. In each case either the outputs remain unchanged or change to all 0s.

Input	Output
$a\ b\ c\ d\ e\ f$	$w\ x\ y$
0 0 0 1 0 1	0 0 0
0 0 1 0 1 0	0 0 0
0 0 0 0 0 1	0 0 0
0 0 0 1 1 1	1 0 1

For Case 2, let us assume input bits a and b change to 1s. The non-code resulting from the first code word covers the third code word. Notice that the third code word in the presence of the unidirectional fault remains unchanged. The non-code word resulting from the second code word does not cover any other code words. The non-code word that is produced from the fourth code word covers all other code words except the second code word. The outputs of the covered code words are either the expected correct value, or there is a single erroneous (0 to 1) transition in the outputs.

5.4   SYNTHESIS OF REDUNDANT FAULT-FREE STATE MACHINES

Redundant faults in a state machine can be eliminated by synthesizing the machine using the procedure suggested in [5.8]. The first step of this procedure uses a *partition-refinement* approach to determine the set of all equivalent state pairs of an S-state sequential circuit. This results in the minimization of states in the original state transition graph such that the reduced graph does not have any equivalent states. For an N-input, S-state sequential circuit, this procedure takes time proportional to $|N|\,|S|\,\log_2 |S|$.

The next step of the synthesis procedure is to assign binary codes to each state in the state transition graph. Many state assignment techniques that generate optimized next state and output logic have been proposed in the literature [5.9, 5.10, 5.11].

Once the states have been encoded, the next state logic and the output logic are synthesized as prime and irredundant combinational logic circuits using the technique proposed in [5.12]. The resulting combinational circuits have $I_i + I_s$ inputs and $I_o + I_s$ outputs, where I_i, I_o, and I_s are the number of primary inputs, primary outputs, and state lines, respectively. The unused states are considered as don't cares during the minimization process.

Let us illustrate the synthesis technique by considering a sequential circuit with I inputs, S states, and O different outputs [5.13]. If the states are encoded using m-out-of-n code such that $\binom{n}{m} \geq S$, then a stuck-at fault in the next state logic, or at the input or output of a memory element, will result in an invalid state. This is because such a fault will result in either a single-bit error or a unidirectional multibit error; in either case a code word will change it into a non-code word, that is, an invalid state. Thus, an invalid state is equivalent to a single or a unidirectional multiple stuck-at fault at the inputs or outputs of memory elements.

Let us assume next that the outputs are also encoded using m-out-of-n code such that $\binom{n}{m} \geq O$. The output logic circuit as shown in Figure 5.10 has two sets of inputs $(y_1, y_2, \ldots, y_n)$ and $(i_1, i_2, \ldots, i_n)$. Set $y_1, y_2, \ldots, y_n$ corresponds to the state variables, and as mentioned above is encoded using m-out-of-n code. Set $i_1, i_2, \ldots, i_n$ corresponds to the primary inputs.

It should be clear from the above discussion that state and output equations are *unate* in state variables and *binate* in primary input variables. These equations can be implemented such that all single stuck-at faults except those at $i_1, i_2, \ldots, i_n$ lines result in single-bit or unidirectional multibit errors irrespective of the way the circuit is implemented. This is because the minimum

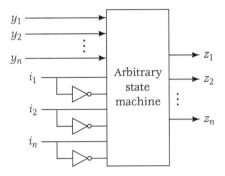

5.10 Output circuit of state machines.

Figure

distance between any two input cubes of such a circuit is at least 2; thus, all bidirectional input cubes are eliminated.

Since a single stuck-at fault in the output logic circuit can cause either a single-bit error or a unidirectional multibit error at the output lines, the output is a non-code word. Similarly, a single stuck-at fault in the next state logic will result in an invalid state, resulting in an invalid output code word. Also, a single stuck-at fault either at an input or at an output of a memory element will force the circuit into an invalid state, thus producing an invalid output. Therefore, in the presence of a single stuck-at fault, the state machine will never produce an erroneous but valid code word; that is, the state machine is fault-secure for any single stuck-at fault.

A state machine that is fault-secure and is also irredundant for all types of faults can be considered totally self-checking [5.13]. For example, let us assume that for a certain input vector, O is the fault-free output of the circuit and O_f the output in the presence of fault f. Because the circuit is irredundant, there must exist an input vector that produces different O and O_f for each fault f in the assumed fault set. Since the circuit is fault-secure, O_f is not a code word if it does not equal O. In other words, the circuit is also self-testing. Therefore, the circuit is both fault-secure and self-testing, and hence totally self-checking.

A state machine, as stated earlier, can have combinational redundant faults and sequential redundant faults. CRFs cannot be detected even if scan flip-flops are used as memory elements in the circuit. However, they can be eliminated if the output and the next state logic are minimized using the external don't cares and *satisfiability* and *observability don't cares* [5.8]. Satisfiability don't cares in a circuit occur as a result of input combinations that never occur. Observability don't cares occur because logic structures at a circuit node may prevent the change in the logic value at the node from being propagated to a primary output. Thus, the logic value at the circuit node is not observable. Since both satisfiability don't cares and observability don't cares arise because of a circuit structure, they are characterized as internal don't cares. Don't cares arising from the operating environment of the circuit rather than the internal circuit structure are called *external* don't cares. Like internal don't cares, external don't cares are composed of satisfiability and observability don't care conditions. For example, if a particular input combination never occurs in a circuit, then this input corresponds to an external satisfiability don't care minterm. Similarly, if an output bit is not used for a set of input patterns, then these patterns form the external observability don't care set of the output. Detailed discussion of the use of don't cares in multilevel logic minimization can be found in [5.8].

The application of the synthesis technique is illustrated next by applying it to a Microelectronics Center of North Carolina (MCNC) benchmark circuit, *bbara* [5.14]. The KISS format of the state machine is shown in Table 5.1(a) [5.15]. The state machine has 4 inputs, 10 states, 50 product terms, and 2 outputs. The first column in the table is the cube corresponding to the primary inputs. The second column is the present state field, the third column the next state field, and the fourth column is the primary output. Since there are 50 product terms, there will be 50 edges in the state transition graph. For example, row 1 in Table 5.1(a) indicates that if the machine is in S0 and the input cube is −0 1, the next state will be S0 and the primary outputs will be 0 0. Table 5.1(b) shows the same state machine with the outputs encoded using 1-out-of-3 code.

The SIS synthesis tool developed at the University of California, Berkeley, was used for state minimization, state assignment, and optimization of next state and output expressions [5.16]. Table 5.1(b) was minimized first by using the STAMINA program in the SIS tool. The state machine was reduced from 10 to 7 states. Next, the 1-out-of-*n* code (*one-hot encoding*) was used for state assignment using the *one-hot* command in SIS. The resulting state expressions are then minimized using the standard minimization script in SIS, and expressed in factored form.

Next, all sequential don't cares (i.e., all unreachable and invalid states) are extracted using the SIS command *extract_ seq_dc*. These don't cares are then utilized, using the command *script.rugged* to further reduce the factored expressions. The resulting next state and output expressions are shown in Table 5.2.

```
.i    4
.o    3
.s    10
.p    60
.r    S0
--01   S0   S0   00        --01   S5   S5   00
--10   S0   S0   00        --10   S5   S5   00
--00   S0   S0   00        --00   S5   S5   00
0011   S0   S0   00        0011   S5   S4   00
-111   S0   S1   00        -111   S5   S1   00
1011   S0   S4   00        1011   S5   S6   00

--01   S1   S1   00        --01   S6   S6   01
--10   S1   S1   00        --10   S6   S6   01
--00   S1   S1   00        --00   S6   S6   01
0011   S1   S0   00        0011   S6   S7   00
-111   S1   S2   00        -111   S6   S1   00
1011   S1   S4   00        1011   S6   S6   01

--01   S2   S2   00        --01   S7   S7   00
--10   S2   S2   00        --10   S7   S7   00
--00   S2   S2   00        --00   S7   S7   00
0011   S2   S1   00        0011   S7   S8   00
-111   S2   S3   00        -111   S7   S1   00
1011   S2   S4   00        1011   S7   S4   00

--01   S3   S3   10        --01   S8   S8   00
--10   S3   S3   10        --10   S7   S8   00
--00   S3   S3   10        --00   S8   S8   00
0011   S3   S7   00        0011   S8   S9   00
-111   S3   S3   10        -111   S8   S1   00
1011   S3   S4   00        1011   S8   S4   00

--01   S4   S4   00        --01   S9   S9   00
--10   S4   S4   00        --10   S9   S9   00
--00   S4   S4   00        --00   S9   S9   00
0011   S4   S0   00        0011   S9   S0   00
-111   S4   S1   00        -111   S9   S4   00
1011   S4   S5   00        1011   S9   S4   00
```

5.1

Table

(a) KISS format of state machine *bbara*.

```
.i        4
.o        3
.s        1 0
.p        60
.r        S0
- - 0 1   S0   S0   1 0 0        - - 0 1   S5   S5   1 0 0
- - 1 0   S0   S0   1 0 0        - - 1 0   S5   S5   1 0 0
- - 0 0   S0   S0   1 0 0        - - 0 0   S5   S5   1 0 0
0 0 1 1   S0   S0   1 0 0        0 0 1 1   S5   S4   1 0 0
- 1 1 1   S0   S1   1 0 0        - 1 1 1   S5   S1   1 0 0
1 0 1 1   S0   S4   1 0 0        1 0 1 1   S5   S6   1 0 0

- - 0 1   S1   S1   1 0 0        - - 0 1   S6   S6   0 0 1
- - 1 0   S1   S1   1 0 0        - - 1 0   S6   S6   0 0 1
- - 0 0   S1   S1   1 0 0        - - 0 0   S6   S6   0 0 1
0 0 1 1   S1   S0   1 0 0        0 0 1 1   S6   S7   1 0 0
- 1 1 1   S1   S2   1 0 0        - 1 1 1   S6   S1   1 0 0
1 0 1 1   S1   S4   1 0 0        1 0 1 1   S6   S6   0 0 1

- - 0 1   S2   S2   1 0 0        - - 0 1   S7   S7   1 0 0
- - 1 0   S2   S2   1 0 0        - - 1 0   S7   S7   1 0 0
- - 0 0   S2   S2   1 0 0        - - 0 0   S7   S7   1 0 0
0 0 1 1   S2   S1   1 0 0        0 0 1 1   S7   S8   1 0 0
- 1 1 1   S2   S3   1 0 0        - 1 1 1   S7   S1   1 0 0
1 0 1 1   S2   S4   1 0 0        1 0 1 1   S7   S4   1 0 0

- - 0 1   S3   S3   0 1 0        - - 0 1   S8   S8   1 0 0
- - 1 0   S3   S3   0 1 0        - - 1 0   S8   S8   1 0 0
- - 0 0   S3   S3   0 1 0        - - 0 0   S8   S8   1 0 0
0 0 1 1   S3   S7   1 0 0        0 0 1 1   S8   S9   1 0 0
- 1 1 1   S3   S3   0 1 0        - 1 1 1   S8   S1   1 0 0
1 0 1 1   S3   S4   1 0 0        1 0 1 1   S8   S4   1 0 0

- - 0 1   S4   S4   1 0 0        - - 0 1   S9   S9   1 0 0
- - 1 0   S4   S4   1 0 0        - - 1 0   S9   S9   1 0 0
- - 0 0   S4   S4   1 0 0        - - 0 0   S9   S9   1 0 0
0 0 1 1   S4   S0   1 0 0        0 0 1 1   S9   S0   1 0 0
- 1 1 1   S4   S1   1 0 0        - 1 1 1   S9   S1   1 0 0
1 0 1 1   S4   S5   1 0 0        1 0 1 1   S9   S4   1 0 0
```

Table 5.1   (b) Output encoded using 1-out-of-3 code.

```
INPUTS = IN_0 IN_1 IN_2 IN_3 q0 q1 q2 q3 q4 q5 q6;
OUTPUTS = Q0 Q1 Q2 Q3 Q4 Q5 Q6 OUT_0 OUT_1 OUT_2;
Node_ 0 = Node_4*Node_5 + !q0*!Node_4 + Node_8 + q3;
Q0 = !Node_ 0;
Node_1 = Node_4*Node_6*!Node_8 + Node_5*Node_8 + !q1*!Node_4 + OUT_2 + !Node_3;
Q1 = !Node_1;
Node_2 = !q3*Node_4*!Node_7 + q3*Node_5 + !q2*!Node_4 + q2*Node_4 + q5;
Q2 = !Node_2;
Q3 = q2*Node_7 + q3*!Node_4;
Node_3 = Node_4*Node_6 + !q4*!Node_4 + !q1*Node_4;
Q4 = !Node_3;
Q5 = q3*Node_7 + OUT_1;
Q6 = !Node_6*Node_8 + OUT_2;
OUT_0 = Node_8 + !Node_3 + Q3 + !Node_2 + !Node_1 + !Node_0 + q3;
OUT_1 = q5*!Node_4 + IN_1*q5;
OUT_2 = q6*!Node_6 + q6*!Node_4;
Node_4 = IN_2*IN_3;
Node_5 = IN_1 + IN_0;
Node_6 = IN_1 + !IN_0;
Node_7 = IN_1*Node_4;
Node_8 = q4*Node_4;
    where Node_0...Node_8 are the internal nodes in the circuit.
```

5.2	Minimized next state and output expressions for *bbara*.
Table	

5.5   DECOMPOSITION OF FINITE STATE MACHINES

The decomposition of a complex state machine into smaller interacting components increases performance and reduces area overhead. Hartmanis [5.17] was the first to propose decomposition of state machines based on preserved partitions. The configuration of a decomposed sequential circuit can be characterized as one of the following [5.18]:

+ Parallel

+ Cascade

+ Arbitrary

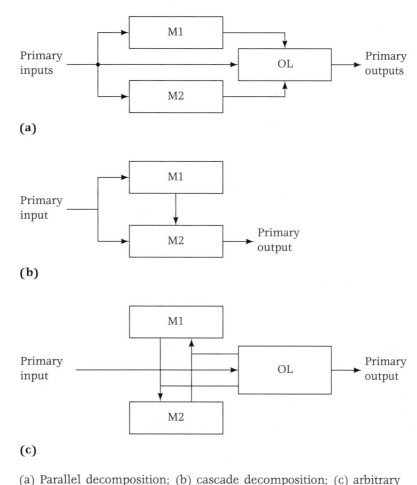

5.11

Figure

(a) Parallel decomposition; (b) cascade decomposition; (c) arbitrary decomposition.

Figure 5.11(a–c) shows parallel, cascade, and arbitrary decomposition, respectively, of a state machine into two components, M1 and M2. In parallel decomposition there is no interaction between components M1 and M2. The components receive the same input sequence via the primary inputs, but they operate independently. The next state lines M1 and M2, as well as the external inputs, feed the output logic block OL for generating the outputs to be observed at primary outputs.

In cascade decomposition, components M1 and M2 are driven by the same inputs but, unlike in parallel decomposition, the components do not operate independently. The next state lines of M1 are direct inputs to M2. Thus, the output of M2, which is also the output of the overall circuit, is controlled by the next state of M1. In this configuration a fault in M1 can be propagated to the primary output via M2. M1 is usually called the *driving* component, and M2 is called the *driven* component.

Arbitrary decomposition allows bidirectional interaction between the component submachines through their state lines. The output logic is driven by the state lines and also by the primary inputs.

To illustrate the decomposition of a state machine, let us consider the state table taken from [5.1] and shown in Figure 5.12(a). This machine is decomposed into three interacting submachines M1, M2, and M3 as shown in Figure 5.12(b) [5.14]. Note that both parallel and cascade decomposition are used in this case. The state tables of submachines M1, M2, and M3 are shown in Figure 5.12(c–e).

5.6   SELF-CHECKING INTERACTING STATE MACHINE DESIGN

Currently available techniques for designing self-checking state machines are in general applicable only to composite circuits. Even if these schemes are used to make all interacting subcircuits of a decomposed circuit self-checking, it cannot be guaranteed that the original circuit itself will be self-checking. This is because of the reduced controllability and observability of the state lines between different submachines. Busaba and Lala [5.13] have proposed a technique for designing interacting finite state machines from the state transition graphs of the constituent submachines; this technique guarantees that the original machine is fault-secure and self-testing, that is, self-checking for all single stuck-at faults.

As stated earlier in Section 5.2, state machines can have combinationally redundant and sequentially redundant faults. If a state machine is configured as a parallel decomposed structure, the redundant faults will be the same as in the original circuit since the submachines operate independently. On the other hand, if a cascade decomposition is used, redundant faults in the resulting configuration can be classified into four categories [5.19]:

1. A fault in M1 that cannot propagate to the state lines.

2. A fault in M1 that propagates to the state lines but not to the primary outputs.

Present state	Input x = 0	x = 1
A	D, 0 1	G, 0 1
B	C, 0 1	E, 0 1
C	H, 1 0	F, 0 1
D	F, 0 1	F, 0 1
E	B, 0 1	B, 0 1
F	G, 0 1	D, 0 1
G	A, 0 1	B, 0 1
H	E, 0 1	C, 0 1

(a)

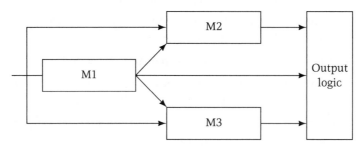

(b)

State	Input x 0	1
P	Q, 1 0	Q, 1 0
Q	P, 0 1	P, 0 1

State	Input xy 0 0	0 1	1 0	1 1
R	R, 1 0	S, 0 1	S, 0 1	S, 0 1
S	S, 0 1	R, 1 0	R, 1 0	R, 1 0

(c) (d)

State	Input xy 0 0	0 1	1 0	1 1
U	V, 0	U, 1	U, 1	V, 1
V	U, 0	V, 1	V, 1	V, 1

(e)

5.12

Figure

(a) State table; (b) implementation of (a) as three interacting submachines; (c) submachine M1; (d) submachine M2; (e) submachine M3.

3. A fault in M2 that does not propagate to primary outputs, but can do so if state lines are completely controllable.

4. A fault in M2 that does not propagate to primary outputs even if state lines are completely controllable.

If none of the submachines has redundant faults, then Types 1 and 4 of redundant faults will not appear in a cascade configuration.

Sequential redundant faults in an arbitrary decomposition can be classified into six categories:

1. A fault in M2 that cannot propagate to its state lines.

2. A fault in M1 that does not propagate to primary outputs, but will do so if state lines of M2 are fully controllable.

3. A fault in M1 that does not propagate to primary outputs even if the state lines of M2 are fully controllable.

4. A fault in M1 that cannot propagate to its state lines.

5. A fault in M2 that does not propagate to primary outputs, but will do so if the state lines of M1 are fully controllable.

6. A fault in M2 that does not propagate to primary outputs even if the state lines of M1 are fully controllable.

Redundant faults of Types 1, 3, 4, and 5 do not occur if none of the submachines has redundant faults in themselves. Redundancies similar to Type 2 in cascade decomposition are not present in the arbitrary decomposition.

A technique for totally self-checking interacting state machine design: A state machine with I inputs and O outputs is first decomposed into two submachines M1 and M2 with S1 and S2 states, respectively. The decomposed machine can be made self-checking if the submachines are fault-secure and self-testing for any single stuck-at faults. This can be guaranteed provided the following conditions are satisfied:

1. m-out-of-n code is used for state assignments such that $\binom{n}{m} \geq S1$ and $\binom{n}{m} \geq S2$.

2. m-out-of-n code is used for output encoding such that $\binom{n}{m} \geq S2$.

3. Next state logic for each submachine and the output logic are implemented such that they are unate in primary input variables and binate in secondary input variables. This guarantees that each submachine and the output logic will be totally self-checking for any single stuck-at fault.

4. Redundant faults are removed from the interacting submachines using the techniques in [5.19].

Let us consider next the impact of the above design technique on each of the previously discussed decomposed configurations.

Parallel decomposition: If both submachines, M1 and M2, are designed by using this technique, the composite machine will be totally self-checking. It should be pointed out that the output logic, which is common to both submachines, produces a non-code word if at any time the state of either M1 or M2 is invalid.

Cascade decomposition: A single stuck-at fault in the next state logic of submachine M1 will create a single-bit error or a unidirectional multibit error at the state lines. Since m-out-of-n code is used for the state assignment, this fault will place M1 in an invalid state. The state lines of M1 are inputs to M2. Invalid inputs to M2 will force it to an invalid state and/or produce invalid output since the state lines of M1 directly feed the output logic. If the fault produces invalid output, then the machine is fault-secure. Also, if the fault moves to an invalid state, the output will be invalid. The effect of a fault on the output logic or on the next state logic of the submachines is as follows:

1. A fault at the input or output of a memory element in M1 will force it to an invalid state; consequently, M2 will generate invalid output.

2. A fault in the output logic will generate either single-bit error or unidirectional multibit error at the output lines. Therefore, the output will be a non-code word.

3. A single stuck-at fault in the next state logic of submachine M2 will generate unidirectional errors at state lines, which in turn will move M2 to an invalid state, thus creating an invalid output.

4. A fault at the input or output of a memory element in M2 will force M2 to an invalid state, thus generating an invalid output.

It should be clear from the above discussion that a state machine implemented in a cascade configuration is fault-secure. If none of the possible redundancies in a cascade configuration is present, then the circuit is self-checking. Type 1 and Type 4 redundancies cannot occur because both M1 and M2 are self-checking. Type 2 redundancy is also not present because if a fault in M1 is propagated to the state lines, the state lines will present a non-code word to the output logic, which in turn will generate a non-code word output. Type 3 redundancy is eliminated by design methods proposed in [5.19]. Also, combinational redundant faults can be eliminated by methods presented in [5.12]. Therefore, a state machine implemented in cascade configuration is fault-secure and irredundant and hence totally self-checking.

Arbitrary decomposition: In a state machine configured using arbitrary decomposition, each submachine communicates with other submachines through its state lines. The state lines of submachines and the primary inputs feed to output logic. An invalid state code at any of the submachines will cause invalid output. A fault in the next state logic of a submachine will cause either single-bit error or unidirectional multibit error. Therefore, such a fault will force the submachine to an invalid state, which in turn will produce an invalid output. Also, a fault at the input or output of a memory element will result in an invalid state, thus resulting in an invalid output. A fault in the output logic will also result in either single-bit error or unidirectional multibit error at the output. Thus, a state machine decomposed using arbitrary decomposition is fault-secure. Such a state machine, as stated previously, can have six types of redundancies. Types 1, 3, 4, and 5 are eliminated since each submachine by itself has no redundant faults. Types 2 and 5 can be eliminated by methods discussed in [5.19]. Therefore, the state machine is fault-secure and irredundant, hence totally self-checking.

5.7   FAIL-SAFE STATE MACHINE DESIGN

The concepts of fail-safe design have been discussed in Chapter 3. The conditions necessary to make a state machine (Mealy type) fail-safe have been proposed in [5.20]. Let S be the set of normal states in a state machine and SR be the set of states that occur under failure conditions. One condition of a fail-safe state machine is

$$S \cap S_R = \varphi \ldots \tag{5.1}$$

The above equation and the definition of S_R require that an erroneous state can only make a transition to another erroneous state when an arbitrary input value is applied.

Let Δ be a completely specified state transition function such as

$$\forall x \in I, \forall q \in S \quad \Delta(x, q) = \delta(x, q) \in S \dots \tag{5.2}$$

where I is the set of inputs and δ is the state transition function $\delta : I \times S \to S$. Assuming Δ^c is a modification of Δ by a failure condition

$$\forall x \in I, \forall q \in S \quad \Delta(x, q) = \delta(x, q) = q' \in S \dots \tag{5.3a}$$

under normal conditions, and

$$\forall x \in I, \forall q_e \in S_R \quad \Delta^c(x, q'_e) = q'_e \in S_R \dots \tag{5.3b}$$

A state machine designed such that it satisfies Equations (5.1), (5.2), and (5.3) is said to be *state fail-safe.*

One approach to make a state machine fail-safe is to use Berger code to encode each state. The first step is to assign a binary code to each state of the machine as in normal state machine design. Every assigned code word is considered as the information bits of a Berger code with which appropriate check bits are concatenated. A code word with information bits equal to all 0s and check bits equal to all 1s, or vice versa, is avoided.

If the next state expressions corresponding to the state variables in the information bits and the check bits are implemented in the sum-of-products and the product-of sums, respectively, then the state machine is fail-safe for unidirectional errors. To illustrate the application of the Berger code in state fail-safe machine design, let us consider the state table of Figure 5.13.

The state assignment of the machine using the Berger code is shown in Figure 5.14; variables y_1, y_2, and y_3 correspond to the information bits, and y_4 and y_5 correspond to the check bits. The transition table resulting from this state assignment is shown in Figure 5.15.

The next state expressions can be obtained directly from the excitation table and are as follows:

$$Y_1 = \overline{x} y_3 y_4 + \overline{x} y_1 y_4 + x y_2 y_4 + x y_2 y_3 y_5 + x y_1 y_4$$
$$Y_2 = \overline{x} y_2 y_4 + x y_3 y_4$$
$$Y_3 = \overline{x}$$
$$Y_4 = (\overline{x} + y_1 + y_2)(x + y_3)(x + y_5)$$
$$Y_5 = \overline{x}(y_1 + y_2 + y_3) y_4$$

An erroneous state transition in a state machine designed using this approach can be detected by a Berger code checker, discussed in Chapter 4. This

	Input	
Present state	x = 0	x = 1
A	E	B
B	C	D
C	A	D
D	E	D
E	A	D
	Next state	

5.13

Figure

State table.

	y_1	y_2	y_3	y_4	y_5
A	0	0	1	1	0
B	0	1	0	1	0
C	0	1	1	0	1
D	1	0	0	1	0
E	1	0	1	0	1

5.14

Figure

State assignment using Berger code.

Present state	x = 0	x = 1
0 0 1 1 0	1 0 1 0 1	0 1 0 1 0
0 1 0 1 0	0 1 1 0 1	1 0 0 1 0
0 1 1 0 1	0 0 1 1 0	1 0 0 1 0
1 0 0 1 0	1 0 1 0 1	1 0 0 1 0
1 0 1 0 1	0 0 1 1 0	1 0 0 1 0

5.15

Figure

Transition table resulting from Berger code assignment.

approach ensures that the state machine is self-checking as far as state transition is concerned; it also has the property that an erroneous state cannot make a transition to a normal (error-free) state.

5.8   REFERENCES

[5.1]   Kohavi, Z. *Switching and Finite Automata Theory.* McGraw-Hill, 1978.

[5.2]   Ghosh, A., S. Devadas, and A. R. Newton. *Sequential Logic Testing and Verification.* Kluwer Academic Publishers, 1992.

[5.3]   Devadas, S., H.-K. T. Ma, A. R. Newton, and A. Sangiovanni-Vincentelli. "Irredundant sequential machines via optimal logic synthesis." *IEEE Trans. CAD* 9 (January 1990): 8–18.

[5.4]   Wakerly, J. *Error Detecting Codes, Self-Checking Circuits and Applications.* Elsevier North-Holland, 1978.

[5.5]   Diaz, M., and P. Azema. "Unified design of self-checking and fail-safe combinational circuits and sequential machines." *IEEE Trans. Computers,* C-28 (March 1979): 276–281.

[5.6]   Ozguner, F. "Design of totally self-checking asynchronous and synchronous sequential machines." *Proc. Int. Symp. Fault-Tolerant Computing,* 1977, 124–129.

[5.7]   Jha, N. K., and S. J. Wang. "Design and synthesis of self-checking VLSI circuits and systems." *IEEE Trans. CAD* 12, no. 6 (June 1993): 878–887.

[5.8]   Hachtel, G. D., and F. Somenzi. *Logic Synthesis and Verification Algorithms.* Kluwer Academic Publishers, 1995 (p. 259).

[5.9]   Devadas, S., H.-K. Ma, A. R. Newton, and A. Sangiovanni-Vincentelli. "MUSTANG: State assignment of finite state machines for optimal and multi-level logic implementation." *IEEE Trans. CAD* 7, no. 12 (December 1988): 1290–1300.

[5.10] Villa, T., and A. Sangiovanni-Vincentelli. "NOVA: State assignment for finite state machines for optimal logic implementation." *IEEE Trans. CAD* 5, no. 9 (September 1990): 905–924.

[5.11] Du, X., G. Hachtel, B. Lin, and A. R. Newton. "MUSE: A multi-level symbolic encoding algorithm for state assignment." *IEEE Trans. CAD* 10, no. 1 (January 1991): 28–38.

[5.12] Bartlett, K., et al. "Multilevel logic minimization using implicit don't cares." *IEEE Trans. CAD* 7 (June 1988): 723–740.

[5.13] Busaba, F., and P. K. Lala. "An approach for self-checking realization of interacting finite state machines." *VLSI Design Journal* 1, no. 4 (1994): 335–343.

[5.14] Busaba, F. "Logic synthesis for self-checking VLSI design." Ph.D thesis, North Carolina State University, 1993.

[5.15] Lisanke, R. "Logic synthesis benchmark circuits." International Workshop on Logic Synthesis, Research Triangle Park, NC, May 1989.

[5.16]   Sentovich, E. M., et al. SIS: A system for sequential circuit synthesis. Electronics Research Laboratory Memorandum No. UCB/ERL M92/41, Univ. of California, Berkeley, May 1992.

[5.17]   Hartmanis, J. "Symbolic analysis of a decomposition of information processing machines." *Information and Control* 3, no. 2 (June 1960): 154–178.

[5.18]   Devadas, S., and A. R. Newton. "Decomposition and factorization of sequential finite state machines." *IEEE Trans. CAD* 8, no. 11 (November 1989): 723–740.

[5.19]   Ashar, P., S. Devadas, and R. R. Newton. "Irredundant interacting state machines via optimal logic synthesis." *IEEE Trans. CAD* 10 (March 1991): 311–325.

[5.20]   Chung, H., and S. Das. "Design of fail-safe sequential machines using separable codes." *IEEE Trans. Computers,* C-27, no. 3 (March 1978): 249–251.

6 Fault-Tolerant Design

T here are two fundamentally different approaches that can be taken to increase the reliability of computing systems. The first approach is called *fault prevention* (also known as *fault intolerance*) and the second, *fault tolerance*. In the traditional fault prevention approach, the objective is to increase the reliability by a priori elimination of all faults [6.1]. Since this is almost impossible to achieve in practice, the goal of fault prevention is to reduce the probability of a system failure to an acceptable low value. In the fault tolerance approach, faults are expected to occur during computation, but their effects are automatically countered by incorporating *redundancy*—that is, additional resources—so that valid computation can continue even in the presence of faults. These resources may include additional hardware (*hardware redundancy*), the addition of redundant information (*information redundancy*), additional software (*software redundancy*), more time (*time redundancy*), or a combination of all these. They are redundant in the sense that they can be omitted from a system without affecting its normal operation.

It should be emphasized that fault tolerance alone cannot enhance the reliability of a system to a desired level. It has to be used in conjunction with the traditional objectives of reliable system design, such as the use of reliable components and easy maintainability. Thus, while fault tolerance can be used to increase significantly the reliability of an already reliable system, it is of little use if the original system is unreliable.

Most of the early work in fault-tolerant system design was motivated by aerospace applications, and in particular by the requirements for computers to be able to operate unattended for long periods of time. While this application is still an important one, fault tolerance is now regarded as a desirable and in some cases an essential feature of a wide range of computing systems, especially in applications where reliability, availability, and safety are of vital importance. For commercial systems, nonredundant (i.e., fault prevention) techniques have been preferred mainly because a redundant design results in increased overhead in terms of area, power consumption, and the like. Reliability is improved by using reliable components, refined interconnections, and so on. However, this approach has limited effectiveness in counteracting faults in hardware and reducing the number of system disruptions. There are an increasing number of applications in which system failures once per day or once per week are not acceptable. Besides, the use of highly skilled systems analysts and service engineers to maintain computer systems often makes the cost of repair and maintenance extremely high. A fault-tolerant design can provide dramatic improvements in system availability and lead to a substantial reduction in maintenance costs as a consequence of fewer system failures [6.2].

6.1   HARDWARE REDUNDANCY

Hardware redundancy is perhaps the most commonly used redundancy and can be employed in several forms. One of these, known as *static redundancy*, achieves fault tolerance without actually detecting any fault. The second form, known as *dynamic redundancy*, has a built-in fault detection mechanism and the capability to recover from the effect of a fault. The third form of hardware redundancy, known as *hybrid redundancy*, utilizes the features of the other two.

6.1.1   Static Redundancy

Static redundancy, also known as *masking redundancy*, uses extra components such that the effect of a faulty component is masked instantaneously. Two important techniques for obtaining fault masking are triple modular redundancy and the use of error correcting codes.

Triple Modular Redundancy

The most general hardware masking technique is *triple modular redundancy* (TMR). The concept of TMR, originally suggested by von Neumann [6.3], is illustrated in Figure 6.1. The boxes labeled M are identical "modules" that feed a *voting element* V (called a "majority organ" by von Neumann). A module may

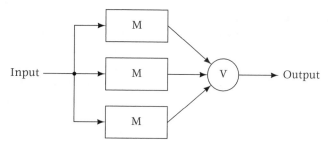

Triple modular redundancy.

be a microprocessor or a less complex unit, such as an adder or a gate. The voting element accepts the outputs from the three sources and delivers the majority vote as its output.

The concept of TMR can be expanded to include any number of redundant modules to produce an N-modular redundancy (NMR) system. An NMR system can tolerate up to n module failures, where $n = (N - 1)/2$. In general, in an NMR system N is considered to be an odd number, but it can also be even; for example, the space shuttle has $N = 4$ in its main computer complex [6.4].

The reliability equation for an NMR system is

$$R_{\text{NMR}} = \sum_{i=0}^{n} \binom{N}{i} \cdot (1 - R_M)^i \cdot R_m^{(N-i)}$$

For the TMR case, $N = 3$ and $n = 1$.

The reliability of the TMR scheme can be determined as a function of the reliability R_M of one module, assuming the voting circuit does not fail. The redundant system will function properly as long as any two modules are operational. It is assumed that failures of the three modules are independent of one another. Hence, the reliability R_{TMR} of the TMR scheme is given by

$$R_{\text{TMR}} = \text{probability of all three modules functioning}$$

$$+ \text{probability of any two modules functioning}$$

$$= R_M^3 + 3R_M^2(1 - R_M)$$

$$= 3R_M^2 - 2R_M^3 \tag{6.1}$$

The reliability of the TMR system is usually better than that suggested by Equation (6.1), since the system may continue to function correctly even if two modules fail. For example, if one module in Figure 6.1 has a stuck-at-1 fault on its output while another module has a stuck-at-0 fault on its output, the system still produces the correct output. Such multiple module failures that do not lead to system failures are termed *compensating module failures* [6.5].

An interesting observation can be made on Equation (6.1). If $R_M = 0.5$, $R_{TMR} = 0.5$, which means there has been no improvement in the overall reliability of the system. This is an example of the general truth that reliability cannot be enhanced if redundancy is applied at a level where the nonredundant reliability is very low.

If it is assumed that each module in a TMR system has passed through an extensive burn-in period, then R_M is an exponential function of time with a constant failure rate λ, or $R_M = e^{-\lambda t}$. Substituting the value of R_M in Equation (6.1), we have

$$R_{TMR} = 3e^{-2\lambda t} - 2e^{-3\lambda t} \tag{6.2}$$

The MTBF (see Section 1.2) of the TMR system is

$$\int_0^\infty R_{TMR} \, dt = \frac{5}{6\lambda}$$

which is less than the MTBF of the individual modules.

Although the MTBF is a useful parameter that is frequently used in establishing the level of reliability in a system, it does not provide much insight into the improvement in reliability given by a fault-tolerant system [6.6]. The basic reason for this is that the MTBF computation evaluates the reliability function for $0 \le t \le \infty$. When redundancy is introduced to improve the reliability of a system, the only region of concern is $0 \le t \le T$, where T is some specified mission time for which the highest reliability is desired; what happens to the system after $t > T$ is not important.

Figure 6.2 shows the plot of reliability functions for $R_{TMR}(t)$ and $R_M(t)$. Although the MTBF $(5/6\lambda)$ of the TMR system is less than that $(1/\lambda)$ of the simplex system, the reliability of the TMR system is much higher for the indicated mission time T.

A parameter that has been found more useful than MTBF for evaluating reliable systems is the *reliability improvement factor* (RIF). It is defined as the ratio of the probability of failure of the nonredundant system to that of the redundant system. If R_N and R_R are the reliabilities of the nonredundant and the redundant systems, respectively, for a fixed mission time T, then

$$RIF = \frac{1 - R_N}{1 - R_R}$$

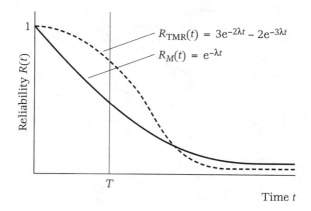

$$R_{\text{TMR}}(t) = 3e^{-2\lambda t} - 2e^{-3\lambda t}$$

$$R_M(t) = e^{-\lambda t}$$

6.2

Figure

Graph of functions $R_{\text{TMR}}(t)$ and $R_M(t)$.

In case a fixed mission time is not specified, the *mission time improvement factor* (MTIF) serves as a convenient comparison measure. It is defined by

$$\text{MTIF} = \frac{T_R}{T_N} \text{ at } R_j$$

where R_j is some predetermined reliability (e.g., 0.99 or 0.90), while T_R and T_N are times at which the system reliabilities $R_R(t)$ and $R_N(t)$, respectively, fall to the value of R_j.

The reliability of the voting element ("voter") was not considered in the reliability expression for TMR. If the voter has the reliability, $e^{-\lambda_1 t}$, Equation (6.2) becomes

$$R_{\text{TMR}} = e^{-\lambda_1 t}(3e^{-2\lambda t} - 2e^{-3\lambda t})$$

This means that the system will fail if the voter fails regardless of whether or not other modules fail. If $\lambda_1 \gg \lambda$, the reliability of the system is less than that of the original system for all values of t. The reliability of a voter in the basic TMR system can be improved by using three identical copies as shown in Figure 6.3. This scheme is called *triplicated TMR*. Two out of the three system outputs are correct if and only if two out of the three replicated voter/module pairs function properly. Hence, the system reliability R_{SYS} is given by

$$R_{\text{SYS}} = (R_M R_V)^3 + 3(R_M R_V)^2(1 - R_M R_V) \tag{6.3}$$

where R_V is the reliability of the voter.

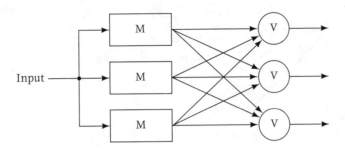

Triplicated TMR system.

If $R_V = 1$, Equations (6.1) and (6.3) become equal, but the basic TMR system is preferable in this case because it uses fewer components.

If a module fails in a TMR system, both the remaining modules must continue to operate correctly. Therefore, once a module has failed, the reliability of the TMR system is lower than that of an individual module. For example, if M1 fails, the reliability of the system becomes

$$\{(\text{reliability of M2}) \times (\text{reliability of M3})\} = R_m^2$$

where the reliability of a module is R_m. It is possible, however, to improve reliability by switching out one of the remaining good modules together with the faulty module and thereby operating the system in a *simplex* (nonredundant) mode. Such a system is known as a *TMR/simplex* system [6.7]. The reliability expression for a TMR/simplex system may be derived from the Markov model of the system shown in Figure 6.4 [6.8]. (See Appendix for a brief introduction to Markov models.) As can be seen, the system begins in state 1, in which all three modules function correctly. If one of the modules fails, this module is discarded together with one of the functional modules, and the system makes a transition to state 2. If the functional module in state 2 fails, the system will move to the failed state 3. Assuming each module in the system has a constant failure rate of λ, the reliability of a module at time is $e^{-\lambda t}$. The unreliability (the probability of failure at $t + dt$) is

$$= 1 - [1 - \lambda dt + (\lambda dt)^2/2! + \dots] \cong 1 - (1 - \lambda dt) = \lambda dt$$

when dt is very small. In Figure 6.4, the transition probability from 1 to 2 is the probability of one of three modules failing in state 1. Thus, the transition probability is $\binom{3}{1}\lambda dt$, or $3\lambda dt$. Similarly, the transition probability from state 2

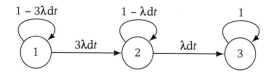

6.4

Figure

Markov model of TMR/simplex system.

to 3 corresponds to the probability of the module failing, that is, λdt. Once the system has entered the failed state 3, it will remain there permanently.

The probability of the system to be at one of the three states at time $(t + dt)$ is as follows:

$$p_1(t + dt) = p_1(t)(1 - 3\lambda dt)$$

$$p_2(t + dt) = p_1(t) \cdot 3\lambda dt + p_2(t) \cdot (1 - \lambda dt)$$

$$p_3(t + dt) = p_2(t) \cdot \lambda dt + p_3(t)$$

The above equations are derived using discrete time and can be represented as

$$\frac{p_1(t + dt) - p_1(t)}{dt} = p_1(t)(-3\lambda)$$

$$\frac{p_2(t + dt) - p_2(t)}{dt} = p_1(t) \cdot 3\lambda + p_2(t) \cdot (-\lambda)$$

$$\frac{p_3(t + dt) - p_3(t)}{dt} = p_2(t) \cdot \lambda$$

The corresponding continuous time equations can be derived by making dt approach 0, which will result in the following differential equations:

$$\frac{dp_1}{dt} = p_1 \cdot (-3\lambda)$$

$$\frac{dp_2}{dt} = p_1 \cdot 3\lambda + p_2 \cdot (-\lambda)$$

$$\frac{dp_3}{dt} = p_2 \cdot \lambda$$

The solution of the differential equations results in

$$p_1(t) = e^{-3\lambda t}$$

$$p_2(t) = 3e^{-\lambda t}/2 - 3e^{-3\lambda t}/2$$

Thus, the reliability of the TMR/simplex system at time t, $R_{\text{sim}}(t)$, is given by

$$R_{\text{sim}}(t) = p_1(t) + p_2(t)$$
$$= 3e^{-\lambda t}/2 - e^{-3\lambda t}/2 = 1.5R_m - 0.5R_m^3$$

One of the major implementation problems associated with an NMR scheme is the synchronization among multiple modules. The most popular approach to the solution of the problem is to use a common clock. However, the clock must be fault tolerant, otherwise there is a possibility that the system may fail due to a fault in the clock. The design of a fully synchronized TMR clock has been described by Davies and Wakerly [6.9]. It uses three modules, each containing a crystal oscillator with feedback through a voter; inputs to the voter are the clock outputs themselves. Lewis [6.10] has described the design of a fault-tolerant clock, which uses the concept of standby spares. Two oscillators, one designated as the primary and the other as the secondary (spare), are used. Initially the primary oscillator is selected; the secondary is switched in if there is a fault in the primary.

Fault detection in a TMR system can be performed by using a set of disagreement detectors, one for each module. A disagreement detector is activated if the output of the module it is attached to is different from that of the voter. The major advantages of the TMR scheme are as follows:

1. The fault-masking action occurs immediately; both temporary and permanent faults are masked.

2. No separate fault detection is necessary before masking.

3. The conversion from a nonredundant system to a TMR system is straightforward.

Triple modular redundancy has been used in Saturn IB and Saturn V on-board computers to increase reliability [6.11]. In the fault-tolerant space computer (FTSC) [6.12], three configuration control units (CCU) have been used in TMR mode for reconfiguring the CPU modules.

6.1.2 Dynamic Redundancy

A system with dynamic redundancy consists of several modules but with only one operating at a time. If a fault is detected in the operating module, it is switched out and replaced by a spare. Thus, dynamic redundancy requires consecutive actions of fault detection and fault recovery. Figure 6.5 illustrates the concept of dynamic redundancy.

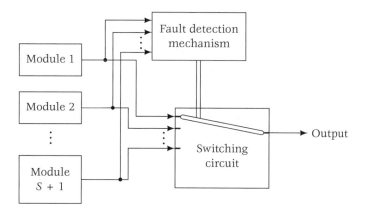

6.5

Figure
Dynamic redundancy scheme with S spares (from [6.13], courtesy of IEEE, ©1975).

A dynamic redundant system with S spares has a reliability

$$R = 1 - (1 - R_m)^{(S+1)}$$

where R_m is the reliability of each module, active or spare, in the system. This reliability function is obtained assuming that the fault detection and the switchover mechanism are perfect. The reliability R is an increasing function of the number of spare modules (Figure 6.6). However, the use of too many spares may have a detrimental effect on the system reliability. Losq [6.13] has shown that for every dynamic redundant system there exists a finite best number of spares for a given mission time. For extremely short mission times, one spare is best. The best number of spares is five or fewer, when the mission time is less than one-tenth of the simplex mean-life.

The detection of a fault in the individual modules of a dynamic system can be achieved by using one of the following techniques:

✦ Periodic tests

✦ Self-checking circuits

✦ Watchdog timers

In periodic tests the normal operation of the functional module is temporarily suspended and a test routine is run to determine if faults are present in the module. A disadvantage of this technique is that it cannot detect temporary faults unless they occur while the module is tested.

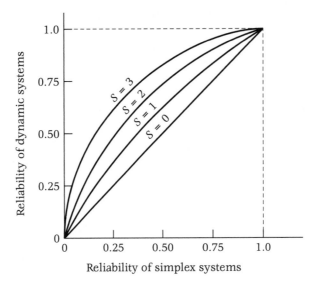

6.6

Figure

Dynamic system reliability as a function of simplex system reliability (from [6.13], courtesy of IEEE, ©1975).

Self-checking circuits provide a very cost-effective method of fault detection (see Chapter 3). They are designed so that for normal circuit inputs they provide correct output or indicate the presence of a fault in a module.

Watchdog timers are an effective and popular method of fault detection. Their principle of operation is relatively simple. Timers are set to certain values at preestablished points, called *checkpoints,* in the program executed by a module. A timer at a particular checkpoint counts down while the module performs its function and is normally reset before the next checkpoint is reached. However, a software bug or a hardware fault will prevent the program from resetting the timer. The timer then issues an interrupt command, which causes automatic switchover to a spare module. The fault recovery process in a dynamic redundant system starts after a fault has been detected in the operating module. It involves switching the faulty module out of service and selecting the system output to come from one of the alternative modules; this process is known as *reconfiguration.*

Before switching out a module from the system, you should determine whether the fault in the module is temporary or permanent; otherwise a good module will be removed because of a temporary fault, which is not economical. A technique that is commonly used for this purpose is called *retry.*

To retry a module, the operation that resulted in the fault is repeated; this requires the knowledge of the module state immediately before the operation was attempted. If the fault is permanent, the retry will be unsuccessful and the defective module must be replaced by a spare that performs the same logic function. In the classical dynamic redundant system, this replacement is invisible to the user, and the system continues its operation uninterrupted; this is known as *self-repair.*

In general, dynamic redundant systems can be divided into two categories:

+ Cold-standby systems

+ Hot-standby systems

In a cold-standby system, one module is powered up and operational, while the rest are not powered; that is, they are "cold" spares. Replacement of a faulty module by a spare is effected by turning off its power and powering a spare. In a hot-standby system, all the modules are powered up and operating simultaneously. If the outputs of all modules are the same, the output of any arbitrarily selected module can be taken as the system output. When a fault is detected in a module, the system is reconfigured so that the system output comes from one of the remaining modules.

The most common arrangement for a hot-standby system is to operate two modules in parallel, with either module acting as a standby. This is known as a *duplex* system. A matching circuit continuously compares the outputs from both modules and interprets any mismatch as a fault in either of the modules or in the matching circuit itself. After the detection of a mismatch, diagnostic programs are run to locate the fault. If the fault is in a module, it is taken off-line and the normal operation is started as a simplex system. When found faulty, the matching circuit gets switched off-line and the reconfigured system returns to operation. The reliability R of a duplex system is

$$R = |R_m^2 + 2CR_m(1 - R_m)|R_S R_c$$

where R_m is the module reliability, R_c the comparator reliability, and R_S the reliability of the module selector circuit. C is the *coverage factor,* which is defined as the conditional probability of successful fault detection and reconfiguration given that a fault exists [6.14]. The reliability of a duplex system is increased if a faulty module is repaired and returned to operation (repairable duplex system). Such a system can be evaluated by Markov modeling. Examples of duplex configurations are the Bell ESS [6.15], the UDET 7116 telephone switching system [6.16], and the COMTRAC railway traffic controller [6.17].

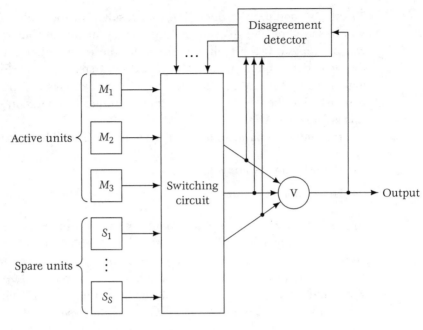

6.7

6.7   Hybrid (3, S) system.

Figure

6.1.3   Hybrid Redundancy

Hybrid redundancy combines the static and the dynamic redundancy approaches. It consists of a TMR system (or in general an NMR system) with a set of spare modules. When one of the TMR modules fails, it is replaced by a spare and the basic TMR operation can continue. The physical realization of a hybrid (3, S) system is shown in Figure 6.7. When a "core" of N active modules is used, it is called the hybrid (N, S) system. The "disagreement detector" detects if the output of any of the three active modules is different from the voter output. If a module output disagrees with the voter, the switching circuit replaces the failed module by a spare. The voter output will be correct as long as at least two modules have correct outputs. If a spare module were to fail in the dormant mode and was switched in on demand from the disagreement detector, this disagreement would still exist and the switching circuit would have to replace it by another module. If all the spares are used up, the system reduces to a hybrid (3, 0) or just a standard TMR.

A hybrid (N, S) system cannot tolerate more than $n| = (N - 1)/2|$ failed modules at a time in the core. For example, if there are two failed modules in the core of a hybrid (3, S) system, the voter will incorrectly switch out the fault-free module and switch a spare in. Since the majority of the modules are malfunctioning, the system will reject all the good spares until all of them are used up and the system will crash.

The reliability of a hybrid system with a TMR core and S spares is

$$R(3, S) = 1 - \{\text{probability of all } (S+3) \text{ modules failing}$$

$$= \quad + \text{ probability of all but one modules failing}\}$$

$$= 1 - \{(1 - R_m)^{(S+3)} + (S+3) \cdot R_m \cdot (1 - R_m)^{(S+2)}\}$$

$$= 1 - (1 - R_m)^{(S+2)}\{1 - R_m + S \cdot R_m + 3R_m\}$$

$$= 1 - (1 - R_m)^{(S+2)} \cdot \{1 + R_m(S+2)\}$$

For systems with standby spares, even though the reliabilities of the spares are higher than those of core modules, it has been assumed in the above reliability expression that they are of equal value, R_m. The voter switch and the disagreement detector (VSD) are assumed to be perfect, for simplicity.

The reliability for a hybrid (N, S) system, where $N = 2n + 1$, can be derived in a similar manner and is equal to

$$R(N, S) = \sum_{i=0}^{i=n+S} \binom{N+S}{i} \cdot (1 - R_m)^i \cdot R_m^{(N+S-i)}$$

assuming no more than n modules fail simultaneously.

The VSD reliability is an important factor in determining the overall system reliability. A reliable switch design for a hybrid (3, 2) system has been described by Siewiorek [6.18]; this is shown in Figure 6.8. The five condition flip-flops (C-FF), referred to in Figure 6.8, record whether a module has disagreed with the voter output. The disagreement detection is accomplished by the EXCLUSIVE-OR of the module and the voter outputs.

The same clock pulse that admits new data to the module flip-flops is used to record disagreement in the C-FFs. The clock pulse activating the C-FFs must be delayed to prevent a flip-flop being set by a transient signal. An iterative cell array determines the first three nonfailed modules (as recorded by the C-FFs) and assigns them, via the interconnection logic, to be voted on. An iterative cell array is a cascade of identical combinational circuits, called *cells* (Figure 6.9). The terminal behavior of a cell is described by means of a *cell table*, which is analogous to the state table of sequential circuits (Table 6.1). Each cell receives as primary input the condition of the ith module (C_i), that is, failed or functional, and the present state denotes the number of modules that have

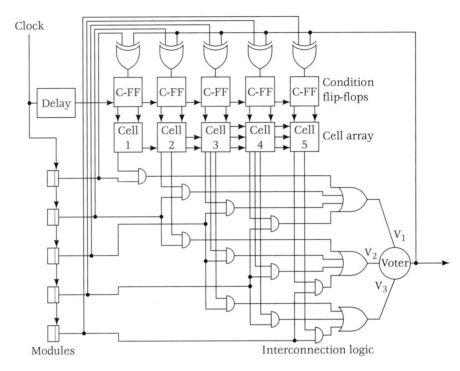

6.8

Figure

An interactive cell switch for a hybrid (3, 2) system (from [6.18], courtesy of IEEE, ©1973).

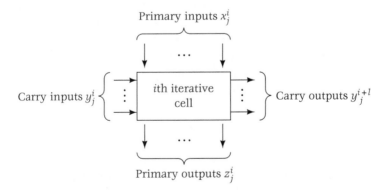

6.9

Figure

An iterative cell (from [6.18], courtesy of IEEE, ©1973).

Present state: S (number of previous modules that are functional)	Next state: S^+		Outputs: $V_1^i V_2^i V_3^i$ *	
	$C_i = 0$ Failed	$C_i = 1$ Functional	$C_i = 1$ Failed	$C_i = 1$ Functional
A (zero)	A	B	0 0 0	1 0 0
B (one)	B	C	0 0 0	0 1 0
C (two)	C	D	0 0 0	0 0 1
D (three +)	D	D	0 0 0	0 0 0

* V_j^i: connect module i to voter input j.

6.1

Table

State and output table for an iterative cell (from [6.18], courtesy of IEEE, ©1973).

been found to be functional prior to the consideration of module i. The output function assigns the particular module to the appropriate voter position.

Next we consider the situation in which module 3 is assumed to have failed, that is, the output of module 3 is different from the voter output. The present state of cell 3 will be C. Since module 3 is faulty, the next state of cell 3 will also be C, and its output will be 0 0 0. Thus, the output of module 3 will be inhibited from appearing at the input of the OR gate driving the V_3 input.

The present state of cell 4 will be C. Since module 4 is functional, the next state of cell 4 will be D and its output will be 0 0 1. Hence the output of module 4 will be transferred to the input of the OR gate connected to the V_3 input of the voter. In other words, module 3, which is faulty, is replaced by module 4.

Thus, the iterative cell switch implements the rotary switching strategy, which requires the least number of switch states. One of the problems associated with the iterative cell switch is that faults can propagate from one cell in the array to the next. Ogus [6.19] has described some techniques to improve the fault tolerance of the VSD using "fail-safe logic" (Chapter 3).

Mathur [6.20] concluded that the reliability of a hybrid (N, 1) system can be improved by keeping $N = 3$ and adding more spares. Cochi [6.21] has shown that the reliability is not always improved by adding spares, because the complexity of the VSD logic increases as more modules are added. He proposed a method for finding an optimum value of spares for which the system reliability is the maximum. It depends upon the reliability of both the VSD and the individual modules. As in dynamic redundant systems, a retry operation can also be used in hybrid systems to recover from temporary faults in modules.

6.2   INFORMATION REDUNDANCY

As indicated earlier, information redundancy involves the addition of redundant information to the original data [6.22]. For example, error detecting and correcting codes append check bits (redundant information) to the data bits to enable detection and correction of erroneous bit(s). This section will illustrate the application of error correcting codes (Chapter 2) in incorporating fault tolerance capability in state machines and in memory systems.

6.2.1   Fault-Tolerant State Machine Design Using Hamming Codes

Armstrong [6.23] has shown how reliable circuits can be constructed using error correcting codes. The technique is in fact a generalization of a triple modular redundancy scheme and is applicable to both combinational and sequential circuits. Design techniques for single-fault tolerant sequential circuits using error correcting codes have also been proposed by Russo [6.24] and Meyer [6.25]. A circuit is single-fault tolerant if it produces correct outputs even in the presence of a single fault. Single-fault tolerance in state machines is achieved by a state assignment with a minimum Hamming distance of 3; the minimum Hamming distance of 3 is required in order to detect and correct a single-bit error. The state assignment for a single-fault tolerant design of the state table for the machine of Figure 6.10(a) is shown in Figure 6.10(b). The state table is then modified so that each state and all its adjacent states are assigned the same next state entry. The modified state table is shown in Figure 6.10(c). The machine continues to operate properly even if a state variable assumes a false value; this is because of the single-error correcting properties of the state assignment. For example, when the input $x = 0$, the next state entry for $0\,1\,1\,1\,1$(C), $1\,1\,1\,1\,1$, $0\,0\,1\,1\,1$, $0\,1\,0\,1\,1$, $0\,1\,1\,0\,1$, and $0\,1\,1\,1\,0$ is $1\,1\,1\,0\,0$, the binary assignment for state B. Thus, the recovery from an erroneous state is possible provided the distance between the correct and the erroneous state is 1. The next state equations for the machine $Y_i(i = 1.5)$ are derived from the modified state table and realized with D flip-flop, by making $D_i = Y_i$. The sharing of logic among the flip-flops is not permitted in order to ensure that a single fault does not affect more than one state variable.

Osman and Weiss [6.26] have shown that a fault-tolerant state machine can be implemented by using three copies of excitation equations and three copies of output expressions to obtain a TMR-like realization. Two copies of the excitation equations and the output expressions employ *shared logic-basis*

Present state	Input x = 0	x = 1
A	C,0	B,0
B	A,1	D,0
C	B,1	A,1
D	D,1	C,0
	Next state, output	

(a)

	$y_1\,y_2\,y_3\,y_4\,y_5$
A	0 0 0 0 0
B	1 1 1 0 0
C	0 1 1 1 1
D	1 0 0 1 1

(b)

Present state	Input x = 0	x = 1
00000		
10000		
01000	01111	11100
00100		
00010		
00001		
11100		
01100		
10100	00000	10011
11000		
11110		
11101		
01111		
11111		
00111	11100	00000
01011		
01101		
01110		
10011		
11011		
10111	10011	01111
10001		
10010		
00011		

(c)

6.10 Figure   (a) State of a sequential machine; (b) state assignment; (c) fault-tolerant state table.

realization of functions and are implemented using a set of modules. The third copy of each equation (excitation and output) can be realized by any technique. Such a state machine can tolerate any single module fault in the shared logic and any fault in the third copy.

Larsen and Reed [6.27] presented a technique for designing fault-tolerant sequential circuits using majority-logic decodable codes. They showed that for a specified ability to tolerate faults, NMR redundancy is more reliable than coding redundancy. However, they indicated that for a fixed complexity, orthogonal (majority-logic decodable) codes provide a greater performance in reliability.

6.2.2   Error Checking and Correction (ECC) in Memory Systems

Semiconductor memory systems are composed of a host of individual RAMs; each RAM is assigned to one bit of a word in memory. The probability of failure of such a system is directly related to the error rate of the individual RAMs and the number of RAMs in the system. RAM chips are subject to two types of errors—*hard* and *soft*.

Hard errors are permanent errors due to physical defects in chips. Three major defects can create hard errors in RAM chips: metallization and bonding failures, oxide defects, and ion contamination [6.28]. Poor metallization can result in open-circuits between device inputs and outputs. Oxide defects may disable a whole chip or just cause a single-bit failure. Ion contamination can induce failure in the row or column decoders. Initial testing and burn-in significantly reduce hard errors in RAMs during system operation.

Soft errors, sometimes referred to as *soft-fails* or *upsets,* are apparently random, nonrecurring changes in memory logic states. For example, a logical 1 may change to a logical 0 or vice versa. Traditional sources of soft errors are power supply noise, board noise, and marginal devices. It has recently been observed that α-particles produced by the radioactive decay of minute quantities of uranium and thorium in the packaging materials can cause bits to flip in dynamic RAMs [6.29]. These bit-flips are soft failures in that they have no permanent effect on a device; the cell that suffers a failure recovers completely when a new bit is written into it and, thereafter, has no greater probability of failure than any other cell in the RAM.

Soft failures in semiconductor memories due to α-particles were not significant until the introduction of 16K and 64K dynamic RAMs. As the dynamic RAM memory cell shrinks to fit more bits per chip, the capacitance of the cell decreases, and that makes the cell more vulnerable to soft errors due to

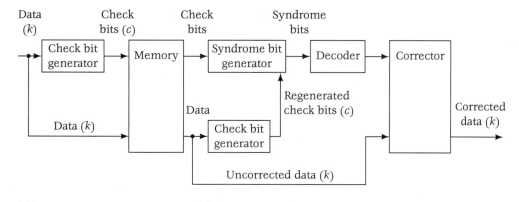

6.11
Figure
A memory system with ECC.

α-particles. Memory manufacturers have taken various precautions to minimize the radiation that hits the chips, but they still get random hits; hence precautions must be taken to make sure these soft errors do not cause system failure.

Errors in semiconductor memory systems are usually accommodated using error checking and correction (ECC) techniques that are based on Hamming or Hsiao codes. Error correction reduces the effect of RAM failures dramatically. Figure 6.11 shows the block diagram of a memory system with error correction. The memory system receives incoming k-bit memory data and generates c check bits, which are stored with the data bits to form a $(k + c)$ memory word. When a memory word is read, the check bits are regenerated from the data bits. Then these check bits are EX-ORed with previously stored check bits to produce *syndrome* (error address) bits. The syndrome bits reveal whether no error occurred, a single-bit error occurred (identifying that bit), or a multibit error occurred. If a single-bit error occurred, a signal is generated that corrects the erroneous bit before it is latched in the output register.

6.2.3 Improvement in Reliability with ECC

The degree of reliability improvement using ECC depends on the failure modes of RAMs. It has been observed that a soft error of a single cell is the dominant failure mode in memory chips [6.30, 6.31]. In fact, if all RAM failures were single-bit failures, they could effectively be eliminated by using ECC. However, RAM devices are subject to several other kinds of hard failures: these can affect a row, a column, or the whole chip. A row failure is caused by a failure of one of the row drivers in a chip, which causes all cells in a row to fail. Similarly, a column failure is caused by a failure of one of the column decoders in a chip, which causes all cells in a column to fail. A whole chip failure causes all the cells in a chip to fail.

ECC memory reliability is also strongly dependent on the reliability of control/support circuits. The reliability of a memory system that incorporates error correction is evaluated below for single-bit failure, whole chip failure, and row (column) failure modes [6.32].

1. *Single-bit failure mode.* Single-bit failures are assumed to be independent events with each cell having a failure rate λ_b and reliability R_b:

$$R_b(t) = e^{-\lambda_b t}$$

Each m-bit word can tolerate a single-bit failure. Thus, the reliability R_{sb} of a given word is

$$R_{sb}(t) = R_b^m + m(1 - R_b) \cdot R_b^{(m-1)}$$

If λ_E is the total failure rate of control/support circuits, then the reliability of the complete W word memory, $R(t)$, is

$$R(t) = e^{-\lambda_E t} \cdot \{R_{sb}(t)\}^W$$
$$= e^{-\lambda_E t} \cdot [e^{-m\lambda_b t} + m \cdot e^{-(m-1)\lambda_b t} - me^{m\lambda_b t}]^W$$
$$= e^{-\lambda_E t} \cdot [me^{-(m-1)\lambda_b t} - (m-1)e^{-m\lambda_b t}]^W \quad \ldots \quad (6.4)$$

2. *Whole chip failure mode.* A single-error correcting system cannot tolerate multiple-bit failures in a word. If a whole chip failure mode is dominant, the memory system must be designed such that each bit in a word belongs to a different chip. Assume a memory system of W words, each word having m bits, is to be implemented with d bit chips. Then the memory system must be organized such that each row has m chips. Since the chips have d locations, each row in the memory system has d words. Thus, the total number of rows in the memory system will be $W/d = \gamma$.

Substituting $\lambda_b = \lambda_C =$ memory chip failure rate, and $W = \gamma$ in Equation (6.4), the reliability $R(t)$ of the memory system becomes

$$R(t) = e^{-\lambda_E t} \cdot [me^{-(m-1)\lambda_C t} - (m-1)e^{-m\lambda_C t}]^\gamma$$

3. *Row (column) failure mode.* If a row (column) failure mode is dominant, the memory system has to be organized such that each bit in a word belongs to a different row (column) in a chip. Thus, even in the presence of a row (column) failure in a chip, each word in the memory system will have a single erroneous bit that can be corrected. In a memory system of W words of m bits each, implemented with d-bit memory chips having q bits per row (column), there will be $p = Wq/d$ sets of rows, each set consisting of m words.

Substituting $\lambda_b = \lambda_\gamma = $ row (column) failure rate, and $W = p$ in Equation (6.4), the reliability of the system is given by

$$R(t) = e^{-\lambda_E t}[me^{-(m-1)\lambda_\gamma t} - (m-1)e^{-m\lambda_\gamma t}]p$$

Maximum gain from an error correcting system can be obtained only if the system is checked periodically for faults that are being corrected. Otherwise the probability of multiple-bit failures in a word increases [6.33]. For example, if in a 22-bit word memory system (16 data bits and 6 check bits), a failure affects all of the memory words, then, after the first failure, the probability of system failure is higher than that of the system without error correction; there are 21 bits that can fail instead of 16. Therefore, to gain maximum benefit from a single-error correcting system, periodic checking should be made to repair hard faults before they become implicated in a double noncorrectable error. The effectiveness of periodic maintenance is dependent on memory size, service intervals, failure distribution, and the like.

6.2.4   Multiple Error Correction Using Orthogonal Latin Squares Configuration

A memory system with built-in single-error correction/double-error detection (SEC/DED) capability can correct any single-bit error in a given word, but fails if any word contains two or more bit errors. Goldberg et al. [6.34] suggested a reconfiguration strategy for correcting multibit errors, by replacing faulty memory chips with spares. The number of additional chips required to implement such a strategy is equal to the number of bit errors the memory system should be able to tolerate. As an alternative to total chip replacement, Hsiao and Bossen [6.35] proposed a scheme that uses the concept of address skewing to dispense multibit errors in a word into several locations, so that the word in each of these locations contains at most a single-bit error. Thus, the error correction capability of the SEC/DED can be utilized despite the presence of multibit errors in a word. The skewing is derived from the theory of *orthogonal Latin squares*.

Let x be a set of n elements $\{x_1, x_2, \ldots, x_n\}$. A Latin square of order n is then an $n \times n$ array of elements of x such that each row and each column of the array contains each element of x exactly once. A pair of Latin squares of order n is said to be orthogonal if, when one square is superimposed on the other, each of the n^2 possible ordered pairs occurs exactly once in the array [6.35]. For example, with $n = 4$, there exist three possible orthogonal Latin squares L_1, L_2, L_3:

$$L_1 = \begin{vmatrix} 0 & 1 & 2 & 3 \\ 1 & 0 & 3 & 2 \\ 2 & 3 & 0 & 1 \\ 3 & 2 & 1 & 0 \end{vmatrix} \quad L_2 = \begin{vmatrix} 0 & 2 & 3 & 1 \\ 1 & 3 & 2 & 0 \\ 2 & 0 & 1 & 3 \\ 3 & 1 & 0 & 2 \end{vmatrix} \quad L_3 = \begin{vmatrix} 0 & 3 & 1 & 2 \\ 1 & 2 & 0 & 3 \\ 2 & 1 & 3 & 0 \\ 3 & 0 & 2 & 1 \end{vmatrix}$$

The result of the superimposition of the first two Latin squares is shown below:

$$\begin{vmatrix} 0,0 & 1,2 & 2,3 & 3,1 \\ 1,1 & 0,3 & 3,2 & 2,0 \\ 2,2 & 3,0 & 0,1 & 1,3 \\ 3,3 & 2,1 & 1,0 & 0,2 \end{vmatrix}$$

Let us now consider how the orthogonal Latin square structure can be built into a 4-word × 4-bit memory system consisting of 4 memory cards, each card having a 4-word × 1-bit memory as shown in Figure 6.12(a).

The L_0 square shown below represents the original address distribution. This is exactly the same as the conventional memory

$$\begin{array}{cccc} 0 & 0 & 0 & 0 \\ 1 & 1 & 1 & 1 \\ 2 & 2 & 2 & 2 \\ 3 & 3 & 3 & 3 \end{array}$$
$$L_0$$

without any reconfiguration capability. The initial state of each linear feedback shift register (LFSR) in the cards is set to $Q_0 Q_1 = 0\,0$. When a card receives the 2-bit address inputs, these bits are EX-ORed with the contents of the LFSR in the card, and the resultant bits are decoded to address the memory chip (Figure 6.12(b)). Thus, if the LFSRs are initially set to all 0s, the memory chips in each card will receive the incoming address. This is exactly the same as the conventional memory without reconfiguration capability.

If a double error is detected, the memory system is reconfigured so that it uses the L_1 configuration. The switchover from the L_0 to L_1 configuration is achieved by setting the states of the LFSRs in cards 2, 3, and 4 to 1 0, 0 1, and 1 1, respectively. Thus, in response to the 2-bit address patterns, the cards receive skewed patterns as shown below:

Address	card1	card2	card3	card4
0 0	0 0(0)	1 0(1)	0 1(2)	1 1(3)
1 0	1 0(1)	0 0(0)	1 1(3)	0 1(2)
0 1	0 1(2)	1 1(3)	0 0(0)	1 0(1)
1 1	1 1(3)	0 1(2)	1 0(1)	0 0(0)

The switching from the L_1 to L_2 configuration can be done by applying the clock pulse once to the boards. This will result in a change of the initial state of the LFSRs in memory cards 2, 3, and 4 to 0 1, 1 1, and 1 0, respectively. The skewed pattern received by the cards corresponding to addressed patterns are as follows:

Address	card1	card2	card3	card4
0 0	0 0	0 1	1 1	1 0
1 0	1 0	1 1	0 1	0 0
0 1	0 1	0 0	1 0	1 1
1 1	1 1	1 0	0 0	0 1

Similarly, the L_3 configuration can be derived from L_2 by clocking the LFSRs once.

For a memory of 2^r addresses there exist $(2^r - 1)$ copies of orthogonal Latin squares. The $(2^r - 1)$ copies of orthogonal Latin squares can be achieved by an r-stage linear feedback shift register characterized by a primitive polynomial of degree r. In general, if a memory has 2^r words each having n bits $(2^r \geq n)$, any k-bit error in a single word, where $1 \leq k \leq n$, can be dispersed into k single-bit errors occurring in k different words using orthogonal Latin squares of order 2^r. As an example, Figure 6.13(a) shows a memory block in which all the bits of the word addressed by #2 are erroneous. In the skewed version (Figure 6.13(b)), using an orthogonal Latin square, the erroneous bits are dispersed so that each addressed word has a single-bit error.

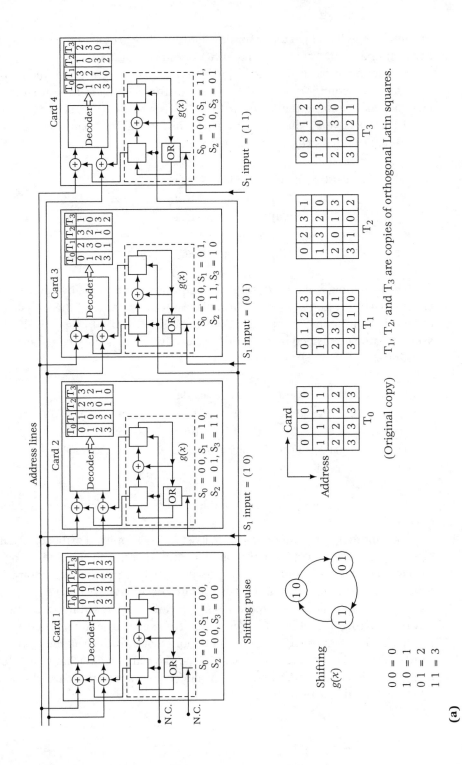

(a) Memory organization showing Latin square configuration. *(Continued on next page)*

6.12
Figure

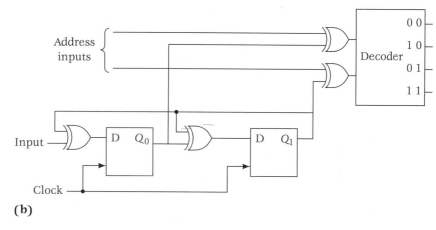

6.12

Figure

(Continued)   (b) Address decoding (from [6.35], courtesy of IEEE, © 1975).

	0	0	0	0	0		0	1	2	3	4
	1	1	1	1	1		2	3	4	0	1
Add #2	2	2	2	2	2		4	0	1	2	3
	3	3	3	3	3		1	2	3	4	0
	4	4	4	4	4		3	4	0	1	2

(a)   **(b)**

6.13

Figure

Skewed memory-address configuration: (a) original address form; (b) Latin square address form.

6.2.5   Soft Error Correction Using the Horizontal and Vertical Parity Method

An alternative to Hamming codes is the horizontal and vertical parity method [6.36], and this can be economically used to detect and correct soft errors in memory systems. It uses a combination of firmware/software and hardware logic for error correction and is less expensive than Hamming codes for smaller memory systems. The memory is partitioned into a number of blocks. A

Bit number

	7	6	5	4	3	2	1	0	
0	1	1	1	1	1	0	1	1	0
1	0	0	1	0	1	1	0	1	1
2	1	0	0	1	1	0	0	0	0
3	0	0	0	0	0	0	0	0	1
4	1	1	1	1	1	1	1	1	1
5	0	0	0	(1)	0	1	1	1	1
6	1	1	1	0	0	1	0	1	0
7	0	0	1	0	0	0	1	1	0
	0	1	1	0	0	0	0	0	

Word number

6.14

Figure

An eight-word memory block.

horizontal parity bit is added to each word of a block in order to detect an erroneous bit in the word. Once the failed word is identified, the CPU initiates a firmware routine that uses the vertical parity word of the block to locate the erroneous bit. The odd vertical parity word for a block is obtained by EX-ORing the words in the block; every memory write operation updates one vertical parity word. When the horizontal parity bit detects an error in a word on a read operation, the firmware recalculates the vertical parity word of that block by EX-ORing all words except the failed word. This newly calculated vertical parity word is EX-ORed with the original vertical parity word of the block; the result produces the correct data to be stored in the failed location. Figure 6.14 shows a memory block of eight words with eight bits per word, although a block can have any number of words. If, for example, bit 4 of word 5 is in error (i.e., changed from 1 to 0), the firmware is started. It saves the current vertical parity word and calculates the EX-OR of all words except word 5. The result 0 1 1 1 0 1 1 1 is EX-ORed with 0 1 1 0 0 0 0 0 to produce 0 0 0 1 0 1 1 1, which is the original content of location 5. The correct word is then rewritten into location 5, and the original vertical parity word is restored. The method can correct all errors detectable by horizontal parity, including multiple errors in a single word. This is an advantage over Hamming codes, which correct only single-bit errors in a memory word. However, unlike Hamming codes, the method can correct only soft errors.

6.3   TIME REDUNDANCY

Time redundancy is commonly used in the detection and correction of errors caused by temporary faults. It involves the repetition, or *rollback*, of instructions [6.37], segments of programs [6.38], or entire programs [6.39], immediately after a fault is detected. The rollback operation requires that a program restart processing from the last checkpoint, where all the information relevant to the successful execution of the program beyond the checkpoint is stored. If a fault is temporary, rolling back the program to a checkpoint should allow successful recovery. However, if the fault is permanent, the fault detection mechanism will be activated again and an alternative recovery method should be attempted. The Tandem computer makes extensive use of checkpoints [6.40].

The selection of the correct number of checkpoints is important. If too many checkpoints are used, the computation time increases since the normal flow of the program is stopped at each checkpoint in order to save state information. On the other hand, if the checkpoints are too far apart, the recovery time will increase since the program will have to be reexecuted from the last checkpoint. The choice of strategic checkpoints is itself a complex problem. Chandy [6.41] has studied several rollback and recovery models and has determined an optimum checkpoint interval that uses the hardware failure rate. O'Brien [6.42] has also studied the strategies for inserting checkpoints.

If a fault occurs during the creation of a checkpoint, it may result in an incorrect resumption of a program. The protection of the rollback mechanism may be provided by saving the checkpoint information in two separate areas [6.43]. When a fault occurs, the program restarts from the area indicated to be valid by a flag. Finally, care must also be taken not to repeat certain events, called *singular events,* which should only be carried out once. The reexecution of such events because of rollback may have serious consequences on a computer-controlled system.

Time redundancy can also be effectively employed to detect faults in digital systems. The *alternating logic design* technique proposed by Reynolds and Metze [6.44] achieves its fault detection capability by utilizing a redundancy in time instead of the conventional space redundancy. It is based on the successive execution of a required function and its self-dual. A function for which the normal output is the complement of the output, when complemented inputs are applied, is known as a self-dual function—for example, a function $f(x)$, where $x = (x_1, x_2, \ldots, x_n)$, is a self-dual if for all x, $f(\overline{x}_1, \overline{x}_2, \ldots, \overline{x}_n) = \overline{f}(x_1, x_2, \ldots, x_n)$. Thus, the principal characteristic of alternating logic is that it provides a true output in one time period and the complementary output in the next time period. The main disadvantage of the technique is that it requires twice as much time to obtain the verified output. Hence it should not be considered for designs for which time is a major factor. However, the increase in hardware it involves is generally modest in comparison with the space redundancy approach (e.g., the duplex system) where two identical functions have to be used for fault detection.

6.4   SOFTWARE REDUNDANCY

It is now widely recognized that even the most thoroughly debugged software still contains faults. In order to improve the reliability of software, provisions may be incorporated to tolerate such faults. Redundancy, used to achieve fault tolerance in hardware, has not found wide application in software. The main problem is that it is not possible to quantify the expected improvement in reliability that can be achieved by using additional software.

Chen and Avizienis [6.45] have suggested the idea of *N-version programming* for providing fault tolerance in software. In concept this approach is similar to the NMR scheme used to provide tolerance against hardware faults. The standard NMR scheme, however, does not provide protection against design faults, since each of the N copies is of identical design. In the N-version programming approach, a number of independently written programs for a given function are run simultaneously; results are obtained by voting upon the outputs from the individual programs. In general the requirement that the individual programs should provide identical outputs is extremely stringent. Therefore, in practice "sufficiently similar" output from each program is regarded as equivalent; however, this increases the complexity of the voters [6.46]. In addition to its ability to tolerate design faults, N-version programming is also capable of masking certain categories of temporary hardware faults [6.47].

Basically, an N-version program incorporates redundancy in design. However, in practice it has been found that as many as 50% of the faults in software for control systems can be attributed to faults in the specification [6.48]. Hence N-versions of a design to the same specification are all likely to be faulty. Randell [6.49] proposed a technique, known as the *recovery block,* for masking software design faults. The recovery block is a program composed of checkpoints, acceptance tests, and alternative procedures for a given task. A typical structure of a recovery block is shown below:

```
ensure      T
by          A
else by     B
else by     C
else error
```

where T is an acceptance test that is evaluated after the execution of the "primary" procedure A, to check that no errors are present; a primary procedure is preceded by the keyword *by*. If the acceptance test is passed, control passes to the next statement following the recovery block. If A fails the acceptance test, the alternate procedure B is entered; an alternate procedure B is tried only after restoring the system state to what it had been before the primary procedure was executed. The acceptance test is repeated to check the successful execution of procedure B. If it fails, procedure C is executed. The alternate procedures are identified by the keywords *else by*. When all alternate procedures are exhausted, the recovery block itself is considered to have failed; the final keywords *else error* emphasize the fact.

6.5   SYSTEM-LEVEL FAULT TOLERANCE

In previous sections we have discussed the basic principle and various types of hardware and software redundancy techniques that may be used to achieve fault tolerance. In practice, fault tolerance in a computing system is considered at the *system level* rather than at the hardware and/or software level. The most widely accepted definition of a fault-tolerant computing system is that *it is a system that has the built-in capability (without external assistance) to preserve the continued correct execution of its programs and input/output functions in the presence of a certain set of operational faults* [6.50]. An operational fault is an unspecified deviation of the correct value of a logic variable in the system hardware or a design fault in the software. Correct execution means that the programs, the data, and the results do not contain errors, and that the execution time does not exceed a specified limit. The types of faults that are encountered during system operation may be categorized into *anticipated* and *unanticipated* faults [6.51].

Anticipated faults are those whose occurrence in an operational system can be foreseen. For example, hardware components in a computing system inevitably deteriorate and give rise to faults. Such faults can be modeled with relative success using single and multiple stuck-at faults, bridging faults, breaks, and the like [6.1]. Unanticipated faults, on the other hand, cannot be foreseen but affect the operation of a system. For example, a VLSI chip can fail in so many modes that it is almost impossible to anticipate the consequences. Another example of faults in this category are design faults, which cannot be predicted except for the general assumption that a complex system is very likely to have such faults. Hence, tolerance of both anticipated and unanticipated faults must be considered in high reliability applications.

In general, a fault-tolerant system must have the following built-in features [6.51]:

+ Error detection mechanism

+ Damage assessment and confinement capability

+ Error recovery capability

The presence of a fault in a system can be detected only if the fault creates error(s) in the system output. Thus, an efficient mechanism for error detection is of paramount importance in a fault-tolerant system. Since the checking of output errors is not dependent on the design and implementation of a system, the system may be assumed to be a black box. The checking should be such that errors are detectable if they result from the presence of any fault from the fault set of which the system is tolerant. Also, the probability of simultaneous failure of the system and the checking part should be minimized. Since it is often not possible to realize an ideal checking mechanism, an *acceptable* one is employed in practice. The objective of an acceptable checking mechanism is to enhance the confidence in the correct operation of a system by ensuring that the majority of likely errors at the system output will be detected.

The detection of an error at the output of a system indicates the presence of a fault in the system; however, during the time a fault occurs and its effect manifests as an error at the system output, other parts of a system may get corrupted, leading to more erroneous outputs in the future. Therefore, after an error has been detected, it is extremely important to assess the extent of the system damage by identifying the boundaries in the system state that contain the effect of the fault.

The goal of the error recovery process is to eliminate the error from the system state in order to prevent it from causing a future system failure. The recovery process can be either of the following:

◆ backward recovery

◆ forward recovery

In the backward recovery process, the system is brought back to the correct state it occupied prior to being corrupted by an error. This involves establishing periodic checkpoints and the then-correct state of a system during the execution of a task. When an error is detected, the system is rolled back to the last correct state identified by the checkpoints. In the forward recovery process, an erroneous state is made error-free by appropriate corrective action. Thus, forward recovery is possible if the effect of a fault is anticipated; however, fixing errors that result from unanticipated reasons may be extremely difficult [6.52]. Both recovery processes will be discussed later in detail.

6.5.1   Byzantine Fault Model

At the circuit level only structural faults (e.g., stuck-at faults) that are assumed to have well-defined effects on outputs are considered. Fault tolerance strategies for such faults focus on masking so that output patterns remain error-free. However, faults at the system level typically change the system behavior in an arbitrary manner. For example, a faulty processor in a distributed system may send conflicting information to different processors in the system. Also, two or more faulty processors may collaborate, inadvertently or maliciously, to corrupt the system. Arbitrary behavior by a faulty system component can be modeled as a *Byzantine fault*.

A major problem in a system that assumes the presence of Byzantine faults is to ensure that each fault-free processor computes an output value consistent with the output values computed by the other fault-free processors in the system. A system that operates satisfactorily in the presence of Byzantine faults is denoted as a *Byzantine resilient system*. Lamport et al. [6.53] proposed a theory, referred to as the *Byzantine General's Problem*, that can be used to reach an agreement among the fault-free processors even if the malfunctioning processors exhibit Byzantine faults.

The Byzantine General's Problem assumes a scenario of a group of generals of the Byzantine army and their troops camping outside an enemy city. The generals communicate with each other via messengers and must agree upon a common battle plan. However, one or more of the generals are traitors and send misleading information to the others. Therefore, the objective is to obtain a consensus among all loyal generals. In a fault-tolerant system the processors correspond to the generals, communication links correspond to the messengers, and the treacherous generals correspond to the faulty processors. The consensus problem can be restated as the problem of achieving agreement among the fault-free processors on the value computed by a particular processor in the system. It is assumed that the effect of a faulty link in a system is no different than that of a faulty processor. In other words, a treacherous messenger can be as dangerous as a treacherous general!

A Byzantine resilient system uses certain message exchange protocols, referred to as *Byzantine agreement protocols,* to reach an agreement in the presence of Byzantine faults in a system [6.54]. A major problem in reaching an agreement on the original value sent by a processor is that the receiving processor also needs the value as received by the other processors in the other system. It has been shown that Byzantine agreement can be guaranteed for at most t faults if the following conditions are satisfied:

+ the system must have at least $(3t + 1)$ processors

+ there must be at least $(2t + 1)$ disjoint communication paths between processors

+ the processors must exchange information at least $(t + 1)$ times

+ all processors must be synchronized within a bounded clock skew

To illustrate let us consider a Byzantine resilient system that can tolerate one fault. This system must have at least four processors interconnected by three disjoint paths. Figure 6.15 shows such a system, in which each processor is connected to another processor directly and via two other processors. Next let us assume that processor C sends a message m to the other processors. If processor C is fault-free, the processors A, B, and D receive three copies of the same message because of the configuration of the communication paths. Since all processors have the same message (i.e., m), the message agreed upon is the one originally sent by processor C. Note that in this example the message exchange requires two rounds; in the first round processor C sends m to the other three processors, and in the second round each processor sends back the message it had received in the first round to all other processors. Each processor then takes the majority of the original message and the messages from the other processors. The resulting message is taken to be the message agreed upon by all processors.

Let us now consider the following situation : C is faulty, and A, B, and D are fault-free. Processor C sends message n to A and message m to B and D.

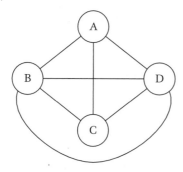

6.15

Figure
A four-processor Byzantine resilient system.

After the second round of message exchange, processors A, B, and D will have three copies of the message as shown below:

	First round	Second round	Message
A	*n*	*m m*	*m*
B	*m*	*m n*	*m*
D	*m*	*m n*	*m*

The majority of three copies (i.e., *m*) are accepted by processors A, B, and D as the agreed message.

The message exchange/voting strategy works even if, instead of C (the originating processor), any other processor is faulty. For example, if B is faulty and sends message *p* to processors A and D in the second round, the three processors will have the following messages:

	First round	Second round	Message
A	*m*	*m p*	*m*
B	*m*	*m m*	*m*
D	*m*	*m p*	*m*

The agreed message in this case is still *m*.

If two processors are faulty in a four-processor system, then the agreement algorithm produces wrong results. As before, let us assume processor C is fault-free and generates message *m*. Suppose processors A and B are faulty and generate message q. In this case after the completion of the message exchange, the processor will have messages as shown below:

	First round	Second round	Message
A	m	m q	m
B	m	m q	m
D	m	q q	q

Since the majority of the three messages processor D received are q, not m, it is not possible to reach an agreement in this case.

6.5.2   System-Level Fault Detection

Fault diagnosis at the system level has the objective of identifying a faulty processor in a multiprocessor system. Since such a system may contain thousands of processors, the identification of a faulty processor in the event of a system failure is a daunting task. In practice a bound on the number of processors that can be faulty is assumed. Preparata et al. [6.55] proposed a fault model for system diagnosis, popularly known as the *PMC model*. This model assumes that a system is composed of n subsystems. Also, it assumes that a fault-free subsystem can comprehensively test another subsystem, and that the results of the tests always accurately identify the correct status of the subsystem. Only permanent faults in a subsystem are considered.

In the PMC model a system is represented as a directed graph with each subsystem being a node. A directed edge connects node i to node j if subsystem i can test subsystem j. The outcome of the test (i, j) is represented by a Boolean variable x_{ij}. If testing node i indicates that node j is fault-free, then $x_{ij} = 0$, whereas if j is faulty then $x_{ij} = 1$. However, if node i is faulty, then the result of the test is unreliable, and x_{ij} can assume any arbitrary value. To illustrate let us consider a four-node system shown in Figure 6.16.

If node a is faulty and node c is fault-free, then $x_{ca} = 1$, but x_{ac} is unpredictable—it may be either 0 or 1. Table 6.2 shows the *syndrome* (the outcome) of the diagnostic tests assuming a single node fault in the system. Note that if the outcome of the test is unpredictable, the corresponding x_{ij} entry in the table is unspecified (represented by a dash). It can be observed from Table 6.2 that a unique binary pattern is associated with a no-fault situation and with each single fault. For example, the pattern 0 0 0 0 indicates no fault, −0 0 1 identifies faulty node a, and so on.

Let us next consider the possibility of two node faults. Table 6.3 shows the corresponding syndrome table. Note that because of unspecified values in

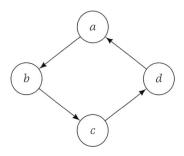

6.16

Figure

A four-node system.

Faulty node	(a,b)	(b,c)	(c,d)	(d,a)
none	0	0	0	0
a	—	0	0	1
b	1	—	0	0
c	0	1	—	0
d	0	0	1	—

6.2

Table

Syndrome for single faulty node.

Faulty node	(a,b)	(b,c)	(c,d)	(d,a)
a,b	—	—	0	1
a,c	—	1	—	1
a,d	—	0	1	—
b,c	1	—	—	0
b,d	1	—	1	—
c,d	0	1	—	—

6.3

Table

Syndrome for two-node faults.

each row, none of the double-node faults can be uniquely identified. Thus, the system of Figure 6.16 is *1-fault diagnosable,* that is, only one node fault can be located.

6.5.3   Backward Recovery Schemes

As mentioned previously, the objective of backward recovery is to make a system return to a point in time during the execution of a task that existed before the system failed. The state of the system at that point, defined as a *checkpoint,* is used to restart the execution of the task. The replacement of the current state of a task by its state at a checkpoint is referred to as *restoration* of the checkpoint [6.51]. The restoration process is initiated once the fault that caused the system failure has been repaired or has vanished automatically (in the case of a transient fault), by *rolling back* the system to the latest checkpoint. Note that the system remains down during the time it takes to roll back and to restore the checkpoint; also, any data received during this period is lost.

A checkpoint is considered to be *active* from the time it is created until it is discarded [6.51]. This period of time is called the *recovery region* or *checkpoint region* of the checkpoint. Appropriate selection of checkpoint locations is extremely important. For example, if a task has to stop frequently for checkpointing, the execution time will be increased. On the other hand, although fewer stops for checkpointing will enable faster execution of a task, the rollback to a checkpoint in case of an error will require going back further in the program. Therefore, the number of instructions for reexecution will be significantly increased. For example, in Figure 6.17 the computation time will be much slower for task B than for task A because of longer rollbacks. In the absence of errors, task B will achieve faster execution. Thus, if the probability of error is high in an operating environment, it is more efficient as far as computation time is concerned to use more checkpoints.

6.5.4   Forward Recovery Schemes

Forward recovery schemes are either hardware-redundancy-based or software-redundancy-based [6.56]. Hardware-based techniques use masking static redundancy to tolerate single faults; however, the actual implementation of masking techniques may vary considerably. For example, Integrity S2 system from Tandem, Inc. uses triplicated processors with voting (based on the TMR concept) to mask single faults [6.57]. Another approach used in the Stratus system [6.58] employs multiple processors to execute the same task. Each processor is made self-checking by duplication, thus enabling it to be *fail-stop.* In other words, if a processor fails—that is, its output does not match with its spare—it

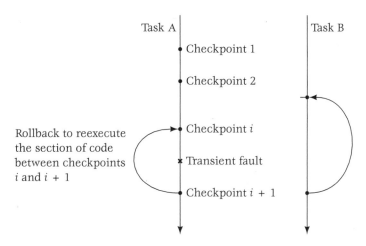

Task A   Task B

Checkpoint 1

Checkpoint 2

Rollback to reexecute
the section of code
between checkpoints
i and $i + 1$

Checkpoint i

Transient fault

Checkpoint $i + 1$

6.17

Figure

Execution of two tasks.

withdraws itself from the system while other processors continue to perform the task. A variation of this approach, known as *reconfigurable duplication* [6.22], uses two processors to execute a task. Thus, a processor failure is indicated if at any stage during the execution of a task, the outputs of the processors do not match. Each processor then runs a diagnostic routine to identify the faulty processor.

A forward recovery scheme based on dynamic redundancy called the *Roll-forward checking scheme* (RFCS) has been proposed by Pradhan and Vaidya [6.59]. This scheme attempts to identify a faulty processor in a duplex system in case of a failure by *retrying* the task on a spare processor. To illustrate the recovery process, let us assume two processors A and B performing the same task as shown in Figure 6.18. As in the backward recovery scheme, the two checkpoints corresponding to the two processors are compared. If they match, the previously stored states of the processors are replaced by the new ones. If the two checkpoints do not match, one of the processors has obviously failed. Assuming processor B has failed during the interval i_j in Figure 6.18, its checkpoint at time t_1 will be different from that of A. This mismatch will activate the recovery mechanism known as *concurrent retry*.

The retry process begins by copying the checkpoint information gathered at time t_0 onto a spare processor S; also, the task assigned to A is transferred to S. Next, S reexecutes the task with the checkpoint information prior to the fault, while A and B continue execution with their respective checkpoints. At time t_2 the checkpoint of S is compared with the mismatching checkpoints of A and B at time t_1. The checkpoint of S will match the checkpoint of A, assuming

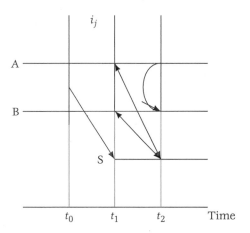

The RFCS scheme.

none of them failed during the period i_j. This matching indicates that B was faulty and A fault-free at time t_1. Then the state of A at time t_2 is copied into B. Therefore, both A and B will be in a correct state if A did not fail in the checkpoint interval i_{j+1}. It should be mentioned that although none of the active processors A and B actually roll back, the recovery process uses *indirect* rollback to retrieve the correct checkpoints by utilizing the spare processor S.

Long et al. [6.60] developed a forward recovery scheme termed the *look-ahead strategy.* In this scheme a task is performed by two processors forming a duplex system. If one of these processors fails, the task is assigned to another two duplex systems. Each of these systems starts execution from one of the mismatching checkpoints of the original duplex system. Also, a separate processor is used to execute the task, starting from the duplex system's checkpoints prior to its failure. At the completion of a checkpoint interval, the processor's checkpoint is compared with the mismatching checkpoints of the original duplex system. This comparison indicates which of these matches with the checkpoint of the processor and hence is correct. The duplex system corresponding to the correct checkpoint then continues with the execution of the task.

6.6 REFERENCES

[6.1]   Lala, Parag K. *Digital Circuit Testing and Testability.* Academic Press, 1997.

[6.2]   Champine, G. A. "What makes a system reliable?" *Datamation,* September 1978, 195–206.

[6.3]   Von Neumann, J. "Probabilistic logics and synthesis of reliable organisms from unreliable components." *Automata Studies,* in Annals of Mathematical Studies no. 34 (Eds.: C. E. Shannon and J. McCarthy), 43–98. Princeton University Press, 1956.

[6.4]   Sklaroff, J. R. "Redundancy management techniques for space shuttle computers." *IBM Jour. Res. & Develop.,* January 1976, 20–27.

[6.5]   Siewiorek, D. "Reliability modelling of compensating module failures in majority voted redundancy." *Proc. Int. Symp. Fault-tolerant Computing,* 1974, 214–219.

[6.6]   Avizienis, A. "Fault tolerance: The survival attribute of digital systems." *Proc. IEEE,* October 1978, 1109–1125.

[6.7]   Ball, M., and H. Hardie. "Majority voter design considerations for TMR computers." *Computer Design,* April 1969, 100–104.

[6.8]   Vainstein, F. Private communication.

[6.9]   Davies, D., and J. F. Wakerly. "Synchronisation and matching in redundant systems." *IEEE Trans. Computers,* C-27 (June 1978): 531–539.

[6.10]   Lewis, D. W. "A fault tolerant clock using stand-by sparing." *Proc. 9th Int. Symp. Fault-Tolerant Computing,* 1979, 33–40.

[6.11]   Dickinson, M. M., J. B. Jackson, and G. C. Randa. "Saturn V launch vehicle digital computer and data adapter." *Proc. Fall Joint Computer Conf.* 26 (1964): 501–516.

[6.12]   FTSC. "The fault-tolerant spaceborne computer." *Digest 6th Int. Symp. Fault-Tolerant Computing,* 1976, 129–147.

[6.13]   Losq, J. "Influence of fault detection and switching mechanisms on reliability of stand-by systems." *Digest 5th Int. Symp. Fault-Tolerant Computing,* 1975, 81–86.

[6.14]   Arnold, T. F. "The concept of coverage and its effect on the reliability model of a repairable system. *IEEE Trans. Computers,* C-22 (March 1973): 251–254.

[6.15]   Toy, W. N. "Fault-tolerant design of local ESS processors." *Proc. IEEE* 66 (October 1978): 1126–1145.

[6.16]   Morganti, M. G., G. Coppadoro, and S. Ceru. "UDET 7116-common control for PCM telephone exchange: Diagnostic software design and availability evaluation." *Digest 9th Int. Symp. Fault-Tolerant Computing,* 1978, 16–23.

[6.17]   Ihara, H., K. Fukuoka, Y. Kubo, and S. Yokota. "Fault-tolerant computer system with three symmetric computers." *Proc. IEEE* 66 (October 1978): 1160–1177.

[6.18]   Siewiorek, D. P., and E. J. McCluskey. "An iterative cell switch design for hybrid redundancy." *IEEE Trans. Computers,* C-22 (March 1973): 290–297.

[6.19]   Ogus, R. C. "Fault-tolerance of the iterative cell array switch for a hybrid redundancy." *IEEE Trans. Computers,* C-23 (July 1974): 667–681.

[6.20]   Mathur, F. P. "On reliability modelling and analysis of ultra-reliable fault-tolerant systems." *IEEE Trans. Computers,* C-20 (November 1971): 1376–1382.

[6.21]   Cochi, B. "Reliability modeling and analysis of hybrid redundancy." *Proc. 5th Int. Symp. Fault-Tolerant Computing,* 1975, 569–577.

[6.22] Johnson, B. W. *Design and Analysis of Fault Tolerant Digital Systems.* Addison-Wesley, 1989.

[6.23] Armstrong, D. B. "A general method of applying error-correction to synchronous digital systems." *Bell Syst. Tech. Jour.* 40 (March 1961): 577–593.

[6.24] Russo, R. L. "Synthesis of error tolerant counters using minimum distance three state assignment." *IEEE Trans. Computers,* EC-14 (June 1965): 359–366.

[6.25] Meyer, J. F. "Fault-tolerant sequential machines." *IEEE Trans. Computers,* C-20 (October 1971): 1167–1177.

[6.26] Osman, M. Y., and C. D. Weiss. "Shared logic realizations of dynamically self checked and fault tolerant logic." *IEEE Trans. Computers,* C-22 (March 1973): 298–306.

[6.27] Larsen, R. W., and I. S. Reed. "Redundancy by coding versus redundancy by replication of failure-tolerant sequential circuits." *IEEE Trans. Computers,* C-21 (February 1972): 130–137.

[6.28] Zachan, M. P. "Error detecting and correcting codes." *Bell Syst. Tech. Jour.* 29 (April 1950): 147–160.

[6.29] Westerfield, E. C. "Memory system strategies for soft errors and hard errors." *Proc. WESCON* 9/1, September 1979, 1–5.

[6.30] Wang, S. Q., and K. Lovelace. "Improvement of memory reliability by single bit error correction. *Proc. COMPCON,* 1977, 175–178.

[6.31] Koppel, R. "RAM reliability in large memory systems—improving MTBF with ECC." *Computer Design,* March 1979, 196–200.

[6.32] Elkind, S. A., and D. P. Siewiorek. "Reliability and performance of error correcting memory and register arrays." *IEEE Trans. Computers,* C-29 (October 1980): 920–926.

[6.33] Rickard, B. "Automatic error correction in memory systems." *Computer Design,* May 1976, 179–182.

[6.34] Goldberg, J., K. N. Levitt, and J. H. Wensley. "An organisation for a highly survivable memory." *IEEE Trans. Computers,* C-23 (July 1974): 693–705.

[6.35] Hsiao, M. Y., and D. C. Bossen. "Orthogonal Latin square configuration for LSI yield and reliability enhancement." *IEEE Trans. Computers,* C-24 (May 1975): 512–516.

[6.36] Edwards, L. "Low cost alternative to Hamming code corrects memory errors." *Computer Design,* July 1981, 143–148.

[6.37] Hopkins, A. L. "A fault-tolerant information processing concept for space vehicles." *IEEE Trans. Computers,* C-20 (November 1971): 1394–1403.

[6.38] Avizienis, A., et al. "The STAR (self testing and repairing) computer: An investigation of the theory and practice of fault-tolerant computer design." *IEEE Trans. Computers,* C-20 (November 1971): 1312–1321.

[6.39] Wensley, J. H., et al. "SIFT: Design and analysis of a fault-tolerant computer for aircraft control." *Proc. IEEE* 66 (October 1978): 1240–1255.

[6.40] Katzman, J. A. "A fault-tolerant computing system." *Proc. 11th Int. Conf. on Syst. Sciences,* Hawaii, January 1978, 85–102.

[6.41] Chandy, K. M. "A survey of analytic models of rollback and recovery strategies." *IEEE Computer,* May 1975, 40–47.

[6.42] O'Brien, F. J. "Rollback point insertion strategies." *Proc. 6th Int. Symp. Fault-Tolerant Computing,* 1976, 138–142.

[6.43] Meraud, C. and P. Lloret. "COPRA, a modular family of reconfigurable computers." *Proc. IEEE Nat. Aerospace and Electronics Conf.,* May 1978, 822–827.

[6.44] Reynolds, D., and G. Metze. "Fault detection capabilities of alternating logic." *IEEE Trans. Computers,* C-27 (December 1978): 1093–1098.

[6.45] Chen, L., and A. Avizienis. "N-version programming: A fault tolerance approach to reliability of software operation." *Digest 8th Int. Symp. Fault-Tolerant Computing,* 1978, 3–9.

[6.46] Randell, B. "Fault tolerant computing system." 6th School of Computing, European Organization for Nuclear Research, September 1980, 362–389.

[6.47] Hecht, H. "Fault tolerant software." *IEEE Trans. Reliability,* August 1979, 227–232.

[6.48] European Workshop of Industrial Computer Systems, TC.7. Hardware of Safe Computer Systems, 1982.

[6.49] Randell, B. "System structure for software fault tolerance." *IEEE Trans. Software Eng.,* SE-1 (June 1975): 220–232.

[6.50] Avizienis, A. "Fault-tolerant systems," *IEEE Trans. Computers,* C-25 (December 1976): 1304–1311.

[6.51] Anderson, A., and P. Lee. *Fault Tolerance: Principles and Practice.* Prentice Hall, 1980.

[6.52] Lee, P. A. "Software faults: The remaining problems in fault tolerant systems." *Hardware and Software Architectures for Fault Tolerance: Experience and Perspectives* (Eds.: M. Banatree and P. A. Lee). Springer-Verlag, 1994.

[6.53] Lamport, L., R. Shostak, and M. Pease. "The Byzantine Generals problems." *ACM Trans. on Programming Languages and Systems* 4 (1982): 382–401.

[6.54] Jalote, P. *Fault-Tolerance in Distributed Systems.* Prentice Hall, 1994.

[6.55] Preparata, F. P., G. Metze, and R. T. Chien. "On the connection assignment problem of diagnosable systems." *IEEE Trans. Electronic Computers,* EC-16 (December 1967): 848–854.

[6.56] Pradhan, D. K. Redundancy Schemes for Recovery. Tech. Report TR-89-CSE-16. ECE Department, University of Massachusetts, 1989.

[6.57] Jewett, D. "A fault-tolerant Unix problem." *Proc. 21st Int. Symp. Fault-Tolerant Computing,* 1991, 512–519.

[6.58] Serlin, O. "Fault-tolerant systems in commercial applications." *IEEE Computer,* August 1984, 19–30.

[6.59] Pradhan, D. K., and N. Vaidya. "Roll forward checkpointing scheme: A novel fault-tolerant architecture." *IEEE Trans. Computers,* C-43 (October 1994): 1163–1174.

[6.60] Long, J., W. K. Fuchs, and J. A. Abraham. "A forward recovery using checkpointing in parallel systems." *Proc. 1990 Int. Conf. on Parallel Processing,* 1990, 402–411.

Appendix:
Markov Models

Markov models are used to analyze probabilistic systems [A.1]. The two key concepts of such models are *state* and *state transition*. A system can be in any one of a finite number of states at any instant of time, and move successively from one state to another as time passes. The changes of state are called *state transitions*. In general, the probability that a system will be in a particular state at time $t + 1$ depends on the state of the system at time $t, t - 1, t - 2$, and so on. However, if the state of the system at time $t + 1$ only depends on the state at time t, and not on the sequence of transitions through which the state at t was arrived at, the system corresponds to a *first-order* Markov model. If the state at time $t + 1$ is independent of the previous state (i.e., the state at time t), then the system corresponds to a Markov model of *zero order*.

The probability that the system will, when in state i, make a transition to state j is known as the *transition probability*. A system with S states has S^2 transition probabilities, which can be denoted by p_{ij}, $1 \le i \le S, 1 \le j \le S$. For computational purposes the transition probabilities are organized into a square matrix P, called the *transition probability matrix*, as shown below; the (i, j) entry in P is the probability of transition from state i to state j.

$$P = \begin{bmatrix} p_{11} & p_{12} & \cdots & p_{1S} \\ p_{21} & p_{22} & \cdots & p_{2S} \\ \vdots & \vdots & & \vdots \\ p_{S1} & p_{S2} & \cdots & p_{SS} \end{bmatrix}$$

In any particular situation, the transition probabilities p_{ij}, and consequently the transition matrix P, depend on what is assumed about the behavior of the system.

The Markov models discussed so far are *discrete time* models. These models require all state transitions to occur at fixed intervals; each transition is assigned with a certain probability. On the other hand, *continuous time* Markov models are characterized by the fact that state transitions can occur at any point in time; the amount of time spent in each state, before proceeding to the next state, is exponentially distributed.

REFERENCE

[A.1]   Shooman, M. L. *Probabilisitic Reliability: An Engineering Approach.* McGraw-Hill, 1968.

Index

1-fault diagnosable system, 196

1-out-of-15 checker, 104–105

1-out-of-n checker, 99, 101–106
  1-out-of-15 checker, 104–105
  1-out-of-n to k-out-of-$2k$ code translator,
    105–106
  design rules, 102–103
  for Mealy-type state machines, 138
  notations and definitions, 101–102
  partitioning the vector space, 103–104
  selection of n code words from k-out-of-$2k$
    code, 103–105

2-bit adder, standard (Berger checker), 115,
  116

2-bit parallel adder (Berger checker), 123, 124

2-out-of-4 checker, 83, 96–98

2-out-of-5 checker, 83, 84–87, 94–95

2-out-of-6 checker, 92–93

3-out-of-6 checker, 98–99

4-out-of-8 checker, 99, 100

5-out-of-10 checker, 99, 100

6-out-of-12 checker, 99, 101

a-bidirectional input cubes, 49

active repair time, 7

alternating logic design technique, 187–188

ALUs, low-cost residue code in, 65

AND gates
  for m-out-of-$(2m + 1)$ checker, 85–87
  in Mealy-type state machines, 140
  in Moore-type state machines, 137–138

AND-OR logic
  in k-out-of-$2k$ checker, 82
  in Mealy-type state machines, 140–141
  in Moore-type state machines, 137

anticipated faults, 189–190

arbitrary decomposition of finite state
    machines, 151, 152, 154, 156

availability
  defined, 8
  dependability and, 12
  repair time and, 8

B1 and B0 encoding schemes for Berger
    codes, 19

backward recovery schemes, 191, 196

bathtub curve, 2, 3

Berger checker, 107–125. *See also* check bit
    generator design (Berger checker);
    secondary group (Berger checker)

Berger codes, 18–19
  B1 and B0 encoding schemes, 19

Berger codes *(cont.)*
    Dong modification, 19
    in fail-safe state machine design, 157–159
    maximal length, 19
    modified, 68, 71
    nonmaximal length, 19
    self-checking checkers for, 107–125
    for totally self-checking PLAs, 68
bidirectional errors
    detection of input fault induced errors,
        53–54
    error-free combinational circuit design,
        50–53
    input encoding approach, 55–58
    output encoding approach, 58–61
bidirectionally adjacent input cubes
    defined, 49
    input encoding for bidirectional error
        elimination, 55–58
    output encoding for bidirectional error
        elimination, 58–61
block code, 15
Boolean functions, self-dual, 62
Borden code, 20
Bose-Lin codes, 20–22
bridging faults in PLAs, 67, 71
burst unidirection error detecting code, 22–27
    Berger codes, 23
    Blaum codes, 24–27
    Bose codes, 23–24
Byzantine fault model, 191–194
    Byzantine agreement protocols, 192
    Byzantine resilient system, 191, 192–193
Byzantine General's Problem, 191–192

cascade decomposition of finite state
    machines, 151–152, 154, 155–156
cellular realization of k-out-of-$2k$ checkers,
    87–90
check bit generator design (Berger checker),
    111–125
    for 8 information bits, 118
    for 11 information bits, 118, 120

    for 15 information bits, 120, 121
    for 16 information bits, 122
    for 32 information bits, 125
check bits. *See also* check bit generator design
    (Berger checker)
    Berger checker check bit generator, 111–125
    in Berger codes, 18–19
    in Bose-Lin codes, 20–22
    in burst unidirectional error detecting
        code, 22–27
    in cyclic codes, 28
    in Hamming code, 31–34
    in Hsiao code, 34–36
    in low-cost residue codes, 27–28
    in parity code, 16
    in strongly fault-secure PLAs, 69
checkpoints
    for backward recovery schemes, 196
    for forward recovery schemes, 197–198
    in look-ahead strategy, 198
    recovery region or checkpoint region, 196
    restoration of, 196
    in RFCS, 197–198
    for rollback, 187, 196
    for watchdog timers, 170
code rate, 15
code words
    in burst unidirection error detecting code,
        23, 24, 27
    covered non-code words and, 48
    covering another word, 17
    covering non-code words and, 48
    in cyclic codes, 28
    defined, 15, 43
    in Hamming code, 33–34
    in input encoding, 57
    in m-out-of-n codes, 18
    for Moore-type state machines, 134–135
    multiple parity bits in, 31
    in Reed-Solomon codes, 39
    selection of n code words from k-out-of-$2k$
        code, 103–105
code-disjoint circuits, 47–48

cold-standby dynamic redundant systems, 171
combinationally redundant faults (CRF), 132, 133, 146–147
concurrent retry, 197–198
condition flip-flops (C-FFs) in hybrid redundant systems, 173
constant failure period (useful life), 2, 3, 4
continuous time Markov models, 204
covering graphs, 50
CRF (combinationally redundant faults), 132, 133, 146–147
cross-point (contact) faults in PLAs, 67, 71
cubes, 49
cyclic codes, 28–29

data polynomials
  cyclic codes corresponding to, 28–29
  Reed-Solomon codes and, 38–39
decoding
  defined, 15
  separable or nonseparable codes and, 16
decomposition of finite state machines, 150–152
  arbitrary decomposition, 151, 152, 154, 156
  cascade decomposition, 151–152, 154, 155–156
  parallel decomposition, 151, 152, 155
dependability, 12
detection of input fault induced bidirectional errors, 53–54
deterministic state machines, 131
disagreement detectors
  in hybrid redundant systems, 172
  in TMR system, 168
discrete time Markov models, 203–204
dynamic redundancy, 168–171. *See also* hybrid redundancy
  cold-standby systems, 171
  fault detection, 168–170
  forward recovery scheme based on, 197–198

hot-standby systems, 171
  reconfiguration, 170–171
  reliability, 169, 170
  retry, 170–171
  watchdog timers, 170

early life (burn-in) period, 2, 3, 4
ECC. *See* error checking and correction (ECC)
equivalent SRF faults, 133–134
error checking and correction (ECC)
  in memory systems, 178–179
  reliability improvement with, 179–181
  row (column) failure mode, 181
  single-bit failure mode, 180
  whole chip failure mode, 180
error correcting codes, 30–40
  Hamming code, 31–34
  Hamming distance and, 30–31
  Hsiao code, 34–36
  multiple parity bits in, 31
  Reed-Solomon codes, 36–40
error correction
  multiple, using orthogonal Latin squares, 181–185
  SEC/DED in memory systems, 181–182
  soft, using parity, 185–186
error detecting codes, 15–29
  Berger codes, 18–19
  Borden code, 20
  Bose-Lin codes, 20–22
  burst unidirection error detecting code, 22–27
  cyclic codes, 28–29
  m-out-of-*n* codes, 18
  multiple error detecting codes, 17–27
  parity code, 16
  residue codes, 27–28
  t-unidirectional error detecting codes, 20–22
  unordered codes for unidirectional error detection, 18–19
exponential failure law, 4

fail-safe combinational circuit design, 73–76
  totally self-checking checker design, 74–76
  ultimate fail-safe systems, 74
fail-safe state machine design, 156–159
  Berger code in, 157–159
  state fail-safe, 156–157
fail-stop in forward recovery schemes, 196
failure rate. *See also* mean-time-between-
      failures (MTBF)
  defined, 2
  exponential failure law, 4
  reliability and, 2–4
  variation with time, 2–3
fault detection. *See also* error detecting codes;
      multiple error detecting codes
  in dynamic redundant system, 168–170
  error detecting codes, 15–29
  input fault induced errors, 53–54
  multiple error detecting codes, 17–27
  SEC/DED in memory systems, 181–182
  system-level, 190, 194–196
  t-unidirectional error detecting codes,
      20–22
  time redundancy and, 187
  in TMR system, 168
fault prevention vs. fault tolerance, 161, 162
fault-secure circuits
  in k-out-of-$2k$ checker, 82
  overview, 44–45
  self-checking checkers and, 79
  in state machines, 134, 135, 146
  strongly fault-secure circuits, 46–47
fault-tolerant design, 161–201. *See also*
      hardware redundancy; information
      redundancy; system-level fault
      tolerance
  fault prevention vs. fault tolerance, 161,
      162
  fault-tolerant system defined, 189
  hardware redundancy, 162–175
  information redundancy, 176–186
  software redundancy, 188–189
  system-level fault tolerance, 189–198

time redundancy, 187–188
faults in state machines, 132–134
  sequentially redundant faults (SRF), 132,
      133–134
  synthesis of redundant fault-free state
      machines, 145–150
finite state machines. *See* state machines
first-order Markov model, 203
forward recovery schemes, 196–198
  hardware-based, 196–198
  look-ahead strategy, 198
  overview, 191
  reconfigurable duplication, 197
  Roll-forward checking scheme (RFCS),
      197–198
  software-based, 196
full adder (Berger checker), 118, 119
fully connected subgraphs, 50
fully specified state machines, 132
functions
  self-dual Boolean, 82
  self-dual complement of, 63
  self-dual, in alternating logic design, 187
  shared logic-basis realization of, 176, 178
  unate, 49, 57

Galois field, 36
gates
  irredundant, 49
  prime, 49
graphs
  covering, 50
  fully connected subgraphs, 50
  subgraphs, 49–50

Hamming code, 31–34
  construction of, 32
  distance-3 code, 33
  distance-4 code, 33–34
  error address generation, 33
  fault-tolerant state machine design using,
      176–178
  Hamming relationship, 32

parity check matrix, 31, 32
Hamming distance of a code
  defined, 17
  error detecting and correcting abilities
    and, 30–31
Hamming relationship, 32
hard errors in memory systems, 178
hardware redundancy, 162–175. *See also*
    dynamic redundancy; hybrid
    redundancy; static redundancy
  dynamic redundancy, 168–171
  in forward recovery schemes, 196–198
  hybrid redundancy, 172–175
  reconfigurable duplication, 197
  static redundancy, 162–168
hardware-based forward recovery schemes,
    196–198
hazard rate. *See* failure rate
horizontal parity, soft error correction using,
    185–186
hot-standby dynamic redundant systems, 171
Hsiao code, 34–36
  check bit generation, 35–36
  double-bit error, 36
  fault-tolerant state machine design using,
    178
  parity check matrix construction, 34–35
  syndrome bits, 36, 179
hybrid redundancy, 172–175
  adding spares, 175
  condition flip-flops (C-FFs), 173
  disagreement detectors, 172
  failed modules tolerated by, 172–173
  iterative cells, 173–175
  NMR system in, 172
  overview, 172
  reliability, 173, 175
  TMR system in, 172
  VSD reliability, 173

incompletely specified state machines, 132
information redundancy, 176–186
  ECC in memory systems, 178–179

fault-tolerant state machine design using
    Hamming codes, 176–178
  multiple error correction using orthogonal
    Latin squares, 181–185
  overview, 176
  reliability improvement with ECC, 179–181
  soft error correction using horizontal and
    vertical parity, 185–186
input code space, 43
input encoding
  for bidirectional error elimination, 55–58
  distance requirements, 55, 57
input fault induced bidirectional errors,
    detecting, 53–54
input space of a circuit, 43
invalid SRF faults, 134
irredundant gate-level circuits, 49
irredundant gates, 49
isomorph SRF faults, 134
iterative cells in hybrid redundant systems,
    173–175

k-out-of-$2k$ checker
  1-out-of-n to k-out-of-$2k$ code translator,
    105–106
  cellular realization of, 87–90
  overview, 82–83
k-out-of-n code in Moore-type state machines,
    135

Latin squares, orthogonal, multiple error
    correction using, 181–185
LFSRs (linear feedback shift registers),
    182–183, 185
linear codes
  cyclic codes, 28–29
  defined, 31
  Hamming code, 31–34
  Hsiao code, 34–36
linear feedback shift registers (LFSRs),
    182–183, 185
literals, 49
look-ahead strategy, 198

low-cost (mod-3) residue codes
  overview, 27–28
  for self-checking design, 65–67
  totally self-checking checker for, 126–127
  for totally self-checking PLAs, 72–73

m-out-of-n checker, 82–106
  for 1-out-of-n code, 99–106
  Berger checker using, 110–111
  cellular realization of, 87–90
  k-out-of-$2k$ checker, 82–83, 87–90
  for m-out-of-$(2m + 1)$ codes, 83, 85–87
  pass transistor-based design for m-out-of-$2m$
    code, 95–99
m-out-of-n codes
  for input encoding, 57
  self-checking checkers for, 82–106
  for self-checking interacting state machine
    design, 154
  for totally self-checking PLAs, 68
maintainability. *See also* mean-time-to-repair
    (MTTR)
  active repair time, 7
  passive repair time, 7
Markov models, 203–204
masking redundancy. *See* static redundancy
maximal length Berger codes
  defined, 19
  totally self-checking checker for, 107–110
Mealy-type state machines. *See also* state
    machines
  AND-OR logic, 140–141
  defined, 131
  encoded state table, 139
  fail-safe design, 156–159
  next state and output equations, 140
  self-checking state machine design,
    138–142
mean-time-between-failures (MTBF). *See also*
    failure rate
  availability and, 8
  as failure rate reciprocal, 5
  reliability, 4–6

  of TMR system, 164
mean-time-to-repair (MTTR). *See also*
    maintainability
  availability, 8
  repair rate, 7
memory systems
  error checking and correction (ECC),
    178–179
  hard errors, 178
  reliability improvement with ECC, 179–181
  soft errors, 178–179
minterms, 49
mission time improvement factor (MTIF) of
    TMR system, 165
mod-3 residue codes. *See* low-cost (mod-3)
    residue codes
modified 2-bit adder (Berger checker), 115,
    116, 124
modified Berger code, 68, 71
modified carry circuit (Berger checker), 117,
    121
Moore-type state machines. *See also* state
    machines
  AND-OR logic, 137
  "divider" of implicants, 137
  on-set realization, 135
  on-set realization of outputs, 136
  output logic checker, 138, 139
  self-checking state machine design,
    134–139
MTBF. *See* mean-time-between-failures
    (MTBF)
MTIF (mission time improvement factor) of
    TMR system, 165
MTTR. *See* mean-time-to-repair (MTTR)
multiple error correction using orthogonal
    Latin squares, 181–185
multiple error detecting codes, 17–27
  Berger codes, 18–19
  Borden code, 20
  Bose-Lin codes, 20–22
  burst unidirection error detecting code,
    22–27

classes of, 17
m-out-of-*n* codes, 18
t-unidirectional error detecting codes,
  20–22
unordered codes for unidirectional error
  detection, 18–19
multiple stuck-at faults, 43–44
multipliers, low-cost residue code in, 65
multiprocessor systems, dependability in,
  12

N-modular redundancy (NMR)
  coding redundancy vs., 178
  in hybrid redundancy, 172
  N-version programming vs., 188
  reliability, 163
  synchronization among multiple modules,
    168
N-version programming, 188–189
next state and output equations
  Mealy-type state machines, 140
  Moore-type state machines, 137
  self-checking interacting state machine
    design, 154
  synthesis of redundant fault-free state
    machines, 145–146
NMR (*N*-modular redundancy)
  coding redundancy vs., 178
  in hybrid redundancy, 172
  N-version programming vs., 188
  reliability, 163
  synchronization among multiple modules,
    168
non-concurrent PLAs, 67
nonmaximal length Berger code, 19
nonseparable codes
  Borden code, 20
  m-out-of-*n* codes, 18

observability don't cares, CRF elimination
  and, 146–147
on-line error detection
  in combinational circuits, 61–62

parity checking to reduce area overhead,
  62
on-set realization
  defined, 135
  of outputs, 136
operational faults, 189
orthogonal Latin squares, 181–185
output code space, 43
output encoding
  for bidirectional error elimination, 58–61
  for Mealy-type state machines, 141, 142
output logic checker for Moore-type state
  machines, 138, 139
output space of a circuit, 43
overlapping parity bits, 16

parallel decomposition of finite state
  machines, 151, 152, 155
parallel systems
  parallel-to-series interconnection, 11–12
  series-to-parallel interconnection, 11–12
parity check matrix
  Hamming code, 31, 32
  Hsiao code, 34–35
  Reed-Solomon code, 39–40
parity checking
  self-dual, 61–65
  soft error correction using, 185–186
parity code, 16
pass transistor-based design for *m*-out-of-2*m*
  checkers, 95–99
passive repair time, 7
periodic tests in dynamic redundant systems,
  169
PLA (programmable logic array) design
  1-out-of-*n* to *k*-out-of-2*k* code translator,
    105–106
  fault types possible, 67
  logic structure, 67, 68
  low-cost residue codes for, 72–73
  modified Berger code for, 68–72
  non-concurrent PLAs, 67
  strongly fault-secure PLAs, 69–72

PLA (programmable logic array) design *(cont.)*
  totally self-checking, 67–73
PMC model for system-level fault detection,
    194–196
prime gates, 49
product lines, 67
product terms for strongly fault-secure PLAs,
    71
programmable logic array design. *See* PLA
    (programmable logic array) design

RAMs. *See* memory systems
reconfiguration, 170–171
recovery block technique, 188–189
redundancy. *See also* dynamic redundancy;
        hardware redundancy; information
        redundancy; static redundancy
  defined, 161
  dynamic redundancy, 168–171
  hardware redundancy, 162–175
  hardware-based forward recovery schemes,
      196–198
  information redundancy, 176–186
  software redundancy, 188–189
  software-based forward recovery schemes,
      197
  static redundancy, 162–168
  time redundancy, 187–188
redundant faults
  combinationally redundant (CRF), 132,
      133, 146–147
  sequentially redundant (SRF), 132,
      133–134
Reed-Solomon codes, 36–40
  constructing, 38–39
  elements of finite fields, 36
  finite fields, 36
  Galois field, 36
  parity check matrix, 39–40
reliability
  defined, 2
  dependability and, 12
  of dynamic redundant systems, 169, 170

ECC for improving, 179–181
exponential failure law, 4
failure rate and, 2–4
fault prevention vs. fault tolerance, 161,
    162
fault tolerance and, 161
of hybrid redundant systems, 173, 175
mean-time-between-failures (MTBF) and,
    4–6
NMR vs. coding redundancy, 178
parallel systems and, 9, 10
parallel-to-series interconnection, 11–12
reliability curve, 5
series systems and, 8, 9–10
series-to-parallel interconnection, 11–12
software redundancy and, 188
of TMR system, 163–164
of TMR system voting element, 165–166
of TMR/simplex system, 168
reliability improvement factor (RIF) of TMR
    system, 164
repair rate
  active repair time, 7
  mean-time-to-repair (MTTR) and, 7
  passive repair time, 7
residue, 27
residue codes
  low-cost (mod-3), 27–28, 65–67
  overview, 27–28
  self-checking design using, 65–67
restoration of the checkpoint, 196
retry
  concurrent, 197–198
  in dynamic redundant systems, 170–171
  in forward recovery schemes, 197–198
RFCS (Roll-forward checking scheme),
    197–198
RIF (reliability improvement factor) of TMR
    system, 164
rollback operation, 187, 196
Roll-forward checking scheme (RFCS),
    197–198
row (column) failure mode in ECC, 181

satisfiability don't cares, CRF elimination
and, 146–147
SEC/DED (single-error correction/double-
error detection) capability in memory
systems, 181–182
secondary group (Berger checker), 115–125
2-bit parallel adder, 123, 124
full adder, 118, 119
modified 2-bit adder, 115, 116, 124
modified carry circuit, 117, 121
overview, 115
standard 2-bit adder, 115, 116
Type I adder, 118, 119
Type II adder, 122, 123
Type III adder, 125
self-checking
circuit properties, 44–46
for single stuck-at faults, 43
unidirectional multiple faults, 43–44
self-checking checkers, 79–129
for Berger codes, 107–125
fault-security and, 79
inputs and outputs, 79, 80
for low-cost residue codes, 126–127
for *m*-out-of-*n* codes, 82–106
in totally fail-safe systems, 74–76
two-rail checker, 79–81
self-checking circuits
for fault detection in dynamic redundancy,
170
fault-secure, 44–45
mod-3 residue code-based, 66–67
self-testing, 44
self-checking combinational logic design,
43–78
bidirectional error elimination techniques,
55–61
bidirectional error-free combinational
circuit design, 50–53
detection of input fault induced
bidirectional errors, 53–54
fail-safe circuit design, 73–76
fault-secure circuits, 44–45

low-cost residue code for, 65–67
on-line error detection, 43
self-checking circuits, 43–45
self-dual parity checking, 61–65
self-testing circuits, 44
strongly code-disjoint circuits, 47–48
strongly fault-secure circuits, 46–47
terminology, 49–50
totally self-checking circuits, 45–46
totally self-checking PLA design, 67–73
self-checking interacting state machine
design, 152–156
arbitrary decomposition and, 154, 156
cascade decomposition and, 152, 154,
155–156
parallel decomposition and, 152, 155
redundant faults, 152, 154
self-checking sequential circuit design,
131–160. *See also* self-checking state
machine design
bidirectional error elimination, 143–144
decomposition of finite state machines,
150–152
fail-safe state machine design, 156–159
faults in state machines, 132–134
self-checking interacting state machine
design, 152–156
self-checking state machine design
techniques, 134–142
synthesis of redundant fault-free state
machines, 145–150
terminology, 131–132
self-checking state machine design, 134–142.
See also Mealy-type state machines;
Moore-type state machines; self-
checking interacting state machine
design
Mealy-type state machines, 138–142
Moore-type state machines, 134–139
on-set realization, 135
on-set realization of outputs, 136
self-checking interacting state machine
design, 152–156

self-dual complement of functions, 63
self-dual functions
    in alternating logic design, 187
    Boolean, 62
    complement of, 63
self-dual parity checking, 61–65
    implementation of, 63–65
    mod-3 residue code for, 66–67
self-testing circuits
    overview, 44
    in state machines, 134
separable codes
    Berger codes, 18–19
    Berger Type I checker, 107
    cyclic codes, 28–29
    residue codes, 27–28
    systematic, 16
separable or nonseparable codes, 16
sequential circuit design, self-checking. *See*
        self-checking sequential circuit design
sequentially redundant faults (SRF)
    equivalent SRF, 133–134
    invalid SRF, 134
    isomorph SRF, 134
    redundant faults, defined, 132
    self-checking interacting state machine
        design and, 152, 154
series systems
    parallel-to-series interconnection, 11–12
    reliability and, 8, 9–10
    series-to-parallel interconnection, 11–12
series-to-parallel interconnection, 11–12
shared logic-basis realization of functions,
        176, 178
single stuck-at faults
    cascade decomposition of state machines
        and, 155
    detecting in combinational logic circuits,
        50
    in k-out-of-$2k$ checker, 82
    in PLAs, 67, 71
    self-checking sequential circuit design and,
        143

single-bit failure mode in ECC, 180
single-error correction/double-error
        detection (SEC/DED) capability
        in memory systems, 181–182
singular events, rollback and, 187
soft errors
    correction using horizontal and vertical
        parity, 185–186
    in memory systems, 178–179
soft-fails. *See* soft errors
software redundancy, 188–189
software-based forward recovery schemes,
        196
SRF (sequentially redundant faults)
    equivalent SRF, 133–134
    invalid SRF, 134
    isomorph SRF, 134
    redundant faults, defined, 132
    self-checking interacting state machine
        design and, 152, 154
state fail-safe, 156–157
state in probabilistic systems, 203
state machines. *See also* Mealy-type state
        machines; Moore-type state machines;
        self-checking state machine design
    decomposition of finite state machines,
        150–152
    deterministic, 131
    fail-safe state design, 156–159
    fault-secure, 134, 135, 146
    fault-tolerant design using Hamming
        codes, 176–178
    faults in, 132–134
    fully specified, 132
    incompletely specified, 132
    Mealy machines, 131, 132
    Moore machines, 131
    redundant fault-free, synthesis of, 145–150
    self-checking design techniques, 134–142
    self-checking interacting design, 152–156
    state transition graph, 131, 132
    states, valid vs. invalid, 131
state transition in probabilistic systems, 203

static redundancy, 162–168. *See also* hybrid
redundancy
    fault detection in TMR system, 168
    mission time improvement factor (MTIF),
        165
    MTBF of TMR system, 164
    N-modular redundancy (NMR), 163, 168
    probability of failed states, 166–168
    reliability, 163–164, 165–166, 168
    reliability improvement factor (RIF),
        164
    TMR/simplex system, 166–168
    triple modular redundancy (TMR),
        162–168
    triplicated TMR, 165–166
    voting element, 162–163, 165–166
strongly code-disjoint circuits, 47–48
strongly fault-secure circuits
    example, 46–47
    for PLAs, 69–72
subgraphs
    defined, 49–50
    fully connected, 50
symmetric errors, 17. *See also* multiple error
    detecting codes
syndrome (error address) bits, 36, 179
system-level fault detection, 194–196
    acceptable checking mechanism, 190
    for one node fault, 194, 195
    PMC model, 194–196
    two node faults and, 194–196
system-level fault tolerance, 189–198
    acceptable checking mechanism, 190
    anticipated faults, 189–190
    backward recovery schemes, 191, 196
    Byzantine fault model, 191–194
    error recover process, 190–191
    forward recovery schemes, 191, 196–198
    operational faults, 189
    system-level fault detection, 190, 194–196
    unanticipated faults, 190
systematic cyclic codes, 28–29
systematic separable codes, 16

t-unidirectional error detecting codes, 20–22
    Borden code, 20
    Bose-Lin codes, 20–22
time redundancy, 187–188
TMR. *See* triple modular redundancy (TMR)
TMR/simplex system, 166–168
    defined, 166
    probability of failed states, 166–167
    reliability, 168
totally fail-safe combinational circuit design,
    73–76
totally self-checking checkers. *See* self-
    checking checkers
totally self-checking circuits
    combinational circuits, 50
    defined, 45
    model of, 45–46
    PLA design, 67–73
transition probability, 203
transition probability matrix, 203
triple modular redundancy (TMR), 162–168
    fault detection, 168
    in hybrid redundancy, 172
    mission time improvement factor (MTIF),
        165
    MTBF, 164
    N-modular redundancy (NMR), 163, 168
    probability of failed states, 166–168
    reliability, 163–164, 165–166, 168
    reliability improvement factor (RIF), 164
    TMR/simplex system, 166–168
    triplicated TMR, 165, 166
    voting element, 162–163, 165–166
triplicated TMR, 165–166
two-rail checker, 79–81
two-rail codes
    self-checking checkers, 79–81
    for totally self-checking PLAs, 68
Type I adder , 118, 119
Type II adder (Berger checker), 122, 123
Type III adder (Berger checker), 125

ultimate fail-safe systems, 74

unanticipated faults, 190
unate functions
    defined, 49
    input encoding and, 57
unidirectional errors. *See also* multiple error
        detecting codes
    Berger codes, 18–19
    Borden code, 20
    Bose-Lin codes, 20–22
    burst error detecting code, 22–27
    m-out-of-n codes, 18
    modified Berger code, 68, 71–72
    in PLA output, 68
    self-checking sequential circuit design and,
        143–144
    t-unidirectional error detecting codes,
        20–22
unidirectional multiple faults, 43–44
unordered codes
    Berger codes, 18–19
    defined, 17
    m-out-of-n codes, 18
    for unidirectional error detection, 18–19

unsymmetric errors, 17
useful life period, 2, 3, 4

vertical parity, soft error correction using,
        185–186
VLSI circuits
    classes of errors in, 17
    totally self-checking PLA design, 67–73
    unidirectional errors in, 17
voting element of TMR system
    overview, 162–163
    reliability of, 165–166

watchdog timers in dynamic redundant
        systems, 170
weakly code-disjoint circuits, 48
wear-out period, 2, 3
whole chip failure mode in ECC, 180
word lines, 67

x-bidirectional input cubes, 49

zero-order Markov model, 203